在极富盛名的威尼斯艺术建筑双年展场，一座造型优美而简约的建筑将作为澳大利亚的代表展现于世人面前。

A sleek and simple
to proudly represe[nt]
the Giardini della B[iennale]
the prestigious Ve[nice]
and Architecture e[vent]

tructure designed
 Australia in
ennale, the heart of
ce Biennale Arts
nts.

威尼斯
建筑双年展
澳大利亚馆
AUSTRALIAN
PAVILION
VENICE

这个两层的展馆占地 320m², 作为一个可灵活布局的艺术陈列空间, 向全世界的参观者展示澳大利亚的视觉艺术及建筑设计作品。| 展馆的设计将简约推向了极致。一个黑色的盒子包裹着一个白色的盒子, 使展馆看上去更像一座雕塑, 而非一座房子。这座展馆的价值不在于建造所用材料, 而是更多体现在设计理念上, 而它所承载的设计理念则需要参观者自己去体会。| 展厅平面是一个简洁的正方形。剖面亦简约, 展馆的主楼层悬于运河之上, 参观者可以由水上进入展馆后方的地面层。四块三米见方的石板像百叶窗一样可以打开, 一块作为门, 其他三块用于采光, 同时可以作为展窗。| 建筑采用钢结构外挂南澳黑色花岗岩石板。展馆内墙为普通白色墙面, 地面采用磨光混凝土。| 在这座展馆中看不到建筑设计的矫揉造作或民族主义的表现。它仿佛是一座茂密花园中的一分子, 既充满力量、自信, 又不失稳重自持。| 澳大利亚馆是一座永久性建筑, 将于 2015 年第 56 届威尼斯双年展期间作为澳大利亚的代表向公众开放。

Within a 320m^2 footprint, the two-level pavilion provides a flexible and adaptable gallery exhibiting Australian visual arts and architecture to international audiences. | The design is of the utmost simplicity, architecturally expressed as a white box contained within a black box. Envisaged as an object rather than a building, it is a container on and in which ideas can be explored, where the container in no way competes with those ideas. | The plan is a simple square. The section is also straight forward with the back-of-house ground level accessible from the water with a piano nobile gallery cantilevered towards the canal. Four 3m stone panels open like shutters: one is the door; others admit light and provide display. | South Australian black granite clads a steel structure. The interior gallery walls are standard white with a polished concrete floor. | Free from architectural affectation and obvious nationalistic statement, it is a powerful, confident yet discreet object within the heavily wooded gardens. | The Pavilion will open in time for the 56th Venice Biennale in 2015 and will remain in the Giardini della Biennale as a permanent structure to represent Australia.

建设地点	意大利 威尼斯
客　　户	澳大利亚艺术协会
荣获奖项	澳大利亚馆设计竞赛头奖 |//

Location Venice, Italy
Client Australia Council for the Arts
Awards First prize Australian Pavilion
shortlisted design competition

TEMPORARY PUBLIC TOILETS RELOCATED WITHIN LANDSCAPE AS PERMANENT STRUCTURE

PAVILION ENTRY APPROACH MORE DIRECT, GENEROUS, ON THROUGH LINK TO BRIDGE, AND SHARED WITH ANOTHER PAVILION

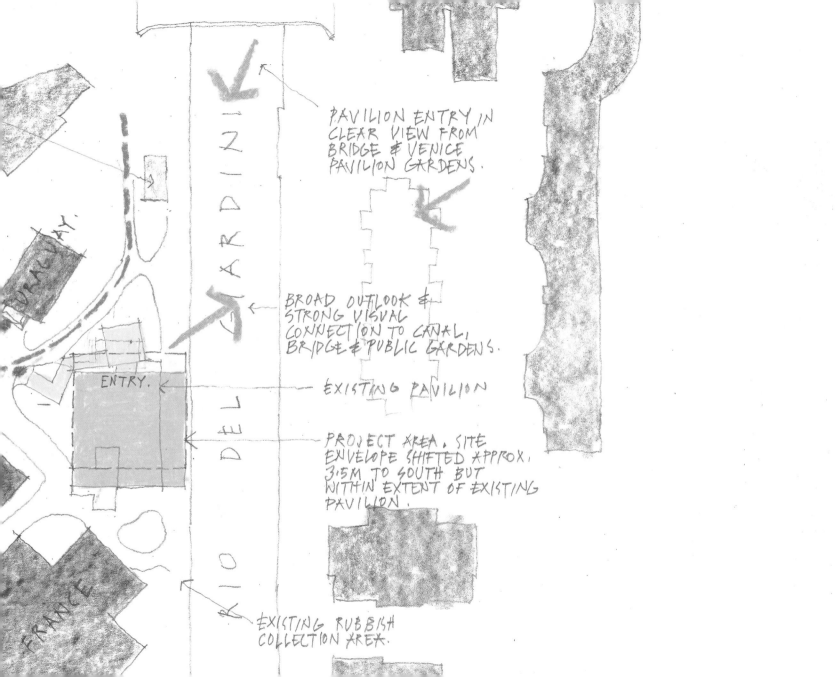

24	前言 —— 支文军教授	
	Foreword by Professor Zhi Wenjun	
32	加速的现代主义 —— 朱剑飞副教授	
	'Modernism Accelerated'	
	by Associate Professor Jianfei Zhu	
90	约翰·丹顿，安尊·菲茨杰拉德访谈录	
	—— 聂建鑫先生	
	Interview with John Denton	
	and Adrian FitzGerald	
	by Jian-Xin Nie	
120	文化 + 公共建筑	
	Cultural + Civic	
246	办公 + 商业	
	Offices + Retail	
304	酒店 + 住宅	
	Hotel + Residential	
434	城市设计 + 城市基础设施	
	Urban Design + Infrastructure	

目录
CONTENTS

前　言

支文军 教授

在我 2002 年初次访问澳大利亚期间，丹顿·廓克·马修建筑设计事务所（DCM）的建筑作品给我留下深刻的印象。该事务所主要代表作之一——墨尔本博物馆是一座身处复杂历史环境中的现代建筑，从设计到建成曾引发异乎寻常的关注和争议，最终依靠实际的城市脉络关系、新旧建筑的整体关系以及建筑自身的艺术魅力，赢得了澳大利亚公众和建筑界的普遍赞赏。我曾撰文《历史对话中的空间塑造——解读墨尔本博物馆》发表在《建筑学报》上。之后的十年间，DCM 的设计动向持续引起我的关注，如在墨尔本的网桥（Webb Bridge）一建成，就在《时代建筑》杂志作了深度报道。

对于中文读者来说，DCM 的社会知名度尚不及活跃在中国一线城市，并逐步向二、三线城市拓进的诸多世界级建筑师及其事务所，其影响力也无法与 DCM 在澳大利亚本土相比，尽管事实上，这已是第二次 DCM 的作品集在中国出版了。回看 DCM 的中国路线，可以有趣地发现，这家成立于 20 世纪 70 年代初，以墨尔本为基地的建筑事务所两次进入中国的经历，似乎是建立在不可预知的偶然之上的必然结果。

30 年前，DCM 承担了它在中国的第一个项目——澳大利亚驻中国大使馆，在城市和建筑尚未进入核心话语的年代，这样一个包裹在政治意义之下的项目在当时显然不会引起人们太多的议论。在这初次接触中，DCM 从类型学方法出发，表达出对中国的初步理解和用现代手法在中国设计建筑的基本理念。之后，直到 21 世纪初，DCM 再次受邀进入中国市场，从商业建筑领域开始建立他们在中国的实践版图，并于 2002 年出版了事务所第一本中文作品集，其中只发表了三个中国项目而且当时都仅为方案阶段。在该版本收录的约翰·丹顿获奖致辞中，他明确提出了 DCM 的"平凡"哲学以及"实用"原则，将来自业主方以及施工方的各种限制归于建筑师职责之内，进而愉快地"出售想法"。如今，DCM 发展出的"草图式"工作框架，可以在短时间内高质量地快速设计出数以万计的空间单元，同时容纳当地的合作与变数。这样一贯的工作态度无疑成为事务所融入中国本土事务的有力保障。

此外，面对中国本土的文化状况，来自澳大利亚的建筑师展现出比西欧或北美的同行更为天然的同理心和相对广泛的接受度。这30年，整个中国在"发展"的呼号声中一路走来，尽管事实上回看每一步都站在十字路口，面对多种可能，需要在极短的时间内作出选择。即使此一时作出了选择，同样的难题又会出现在下一个路口，周而复始。同样的，澳大利亚也始终身处城市与丛林、移民与乡土、亚洲与欧洲的宿命般的文化矛盾与焦虑之中。如何走出一条道路，令丰富而充满想象的城市场景与广袤而真实的大地景观终能汇合在一起，规划者和建筑师心目中的秩序载体得以与空间主体共同创造出城市动力，在这个问题上，澳大利亚的建筑师先于我们很早就开始过各种尝试。

在今天，这些尝试尤其富有记录的意义，因为在 DCM 那里，中国已经投身其中。而这次，同济大学出版社以专题图书的形式系统地介绍 DCM 近十年来的新作及其研究，是值得期待的。

支文军 教授
支文军是同济大学建筑与城市规划学院教授和博士生导师，《时代建筑》杂志主编，同时担任同济大学出版社社长。

FOREWORD

PROFESSOR ZHI WENJUN

During my first visit to Australia in 2002, Denton Corker Marshall's (DCM) works left a great impression on me. Melbourne Museum – one of the practice's major works – is an example of modern architecture placed in a particular historical context. From the initial design right through to the project completion, it attracted wide public attention and controversy. Through the study of the urban context, integration of the old and the new, and demonstration of the aesthetic value that is intrinsic to the architecture, the building eventually won the admiration of the public and the architecture industry. Upon my return, I wrote a paper titled *Spatial Creation along the Historical Context - A Rational Analysis of the Melbourne Museum*, which was published in *Architectural Journal*. In the following 10 years, I continued to observe DCM's design direction and published an in-depth report on Webb Bridge in *Time + Architecture* magazine, upon the project's completion.

Compared to some other international starchitects who are active in first-tier cities in China and gradually expanding to second and third-tier cities, DCM is relatively less well-known to the Chinese market. The practice is considerably less influential in China than in Australia, however, this is the second time that a DCM monograph has been published in China. Interestingly, looking back at DCM's experience in China, we discover that this Melbourne-based practice has entered China twice in history, both seemingly unpredictable but nevertheless inevitable.

Thirty years ago, DCM undertook its very first project in China – the Australian Embassy in Beijing. In an era where urban and architectural design were still absent from the key discourse, a design that was hidden behind political significance, was not destined to draw much attention. In this first attempt, DCM employed the idea of typology and demonstrated its initial understanding of China, and some of the basic concepts of modern design in China. At the beginning of the 21st century, DCM was invited to China for the second time, and was starting to offer their services in commercial development. Three unbuilt projects designed during this period were featured in the first DCM monograph published in China. In John Denton's A. S. Hook Address featured in this monograph, he clearly proposed the philosophical position of "taking the ordinary and trying to make it extraordinary"; and "embracing the realities". This approach incorporates every restraint proposed by developers and builders, and allows architects to "happily sell their ideas". Today, DCM has developed a framework approach referred to as "sketch", which allows the practice to successfully design a huge quantity of housing units with both quality and speed, while also allowing for the contribution from local design institutes and for tight control over design quality. This consistent approach ensures that the practice can integrate successfully into the architecture industry in China.

Moreover, when it comes to the cultural differences, by nature Australian architects are more empathetic than their European or Northern American counterparts. In the last 30 years China has experienced unprecedented growth; with every step forward, there were always decisions to be made in an extremely short time frame. Harder decisions were awaiting at the next phase; and it continued full circle. Similarly, Australia has also always been in the middle of the cultural clash between urban and rural, immigrants and natives, and Asia and Europe. How to integrate the dynamic urban scene with the vast landscape, has become a driving force for Australian architects seeking innovative solutions to create their own identities; and they started experimenting much earlier than us.

Nowadays, these experiments are exceptionally impressive because China plays an important role in DCM's development. Tongji University Press presents a monograph featuring a survey into the practice's projects over the last decade; and it's certainly a great achievement.

Professor **Zhi Wenjun**
Zhi Wenjun is Professor of the College of Architecture and Urban Planning at Tongji University, the Editor-in-Chief of Time + Architecture Journal and the President of Tongji University Press.

论 文

加速的现代主义：
二十一世纪 DCM 的建筑设计
以及中国作为工地
和催化剂的作用与意义

朱剑飞

Denton Corker Marshall（DCM）是当今澳大利亚最杰出的建筑设计事务所之一。DCM 成立于 1972 年，总部设于墨尔本，分公司设于伦敦和雅加达。成立至今，DCM 创造了许多备受赞扬的建筑作品，这些作品不仅分布在澳大利亚，而且在世界各地，包括英国、中国、欧洲大陆和东南亚地区均有公司业绩。虽然这些作品为人所熟悉，并且大量收录在 20 世纪 80 年代后期、2000 年和 2008 年出版的书刊中，[1] 但是 DCM 仍在不断地创作新作品，并给人们带来新的惊喜。由于历史观察距离的局限，这些近年出版的刊物和研究，尚未全面展示 DCM 在 21 世纪 00 年代的发展。从今天——2013 年的历史距离来回顾，我们可以看到，就 DCM 的设计工作而言，21 世纪第一个十年不仅仅是 20 世纪 90 年代的一个简单延续，而是一个呈现出新构思、新观点的新阶段。

比如，在 21 世纪初，DCM 开始在中国参与大型建设项目，在十多个城市中负责设计住宅群体及相关建筑。这些中国项目与同时间在英国设计和兴建的项目形成强烈对比。同时，通过全球化的实践，DCM 的设计体现了对地貌和对世界的一种新的感受，以及对澳洲本土新的理解。这不仅反映在 DCM 在澳大利亚维多利亚州乡间设计的一系列别墅中，亦可以在澳大利亚驻雅加达大使馆和威尼斯双年展澳大利亚馆的设计中得到印证。在 2013 年的今天，关于 DCM 的研究，必须观察探讨这些近年的突破，包括对空间、土地和地缘关系的一种新意识及相关的对于世界的新认识，也包括事务所在不同国家和地区的设计工作，尤其是中国，在 DCM 实践系谱中位于比较突出而特殊的地位。

究竟 DCM 在 21 世纪 00 年代中有何发展？具体发生了些什么？哪些特征标志了 DCM 在这阶段的主要贡献，而不是——90 年代的简单延伸？就 DCM 整体设计工作而言，中国作为一个建筑的基地和市场，究竟处于一个什么位置？DCM 在中国完成了哪些工作？为中国带来了什么？设计建造的最终完成，需要什么样的工作模式与合作方法？市场和设计人之间如何彼此互动？DCM 的贡献在哪里？中国在此发挥了什么样的催化剂作用（如果有这种作用的话）？本文试图回答这些问题，借此了解在 21 世纪的第一个十年 DCM 的整体贡献以及中国作为工地和催化剂对其发展的作用与意义。

1 轮廓

从今天的角度来回顾，DCM 的发展明显包括了三个阶段：首先是 20 世纪 70 年代至 80 年代的"类型学"阶段，这是一个在后现代框架中带有晚期现代思想的阶段（如对抽象的平板和形体的运用）；之后是 90 年代的"至上主义"

阶段，这与当时流行的解构主义思想并行；到了 21 世纪 00 年代则开始了"全球主义"的新阶段，带有新世纪的快速运行的基本特点。当然这些阶段之间的划分不是绝对黑白分明的，很多思路从过去的实践延展到新的框架中，其中既有构思的累积也有对一些以前元素的摒弃。就整体而言，虽然 80 年代后期和 90 年代初期那段时间表现出对其之前作品方向的决定性背离，并形成一种 DCM 独有的形式语言，然而，00 年代并非是单纯的延续，而是明显的升华，其中包括承载压力和效率要求，进行新的实验、发展新的构思。在本文中，我将第三个阶段称为"全球主义"，这并非因为之前缺乏国际性的作品，而是因为 DCM 近年来更加明显地具有以墨尔本为基地向外发展的特征——日益增多的海外项目，同样迅速提高的全球意识和新的地缘和地景观念，以及工作生活中更便捷的通讯交流方式的使用。

在这三阶段的发展中，DCM 的中国实践发生于 1982 年和 2000 年以后。具体来说，DCM 先后两次进入中国：第一次在 1982 至 1992 年期间，设计了澳大利亚驻中国大使馆；第二次从 2000 年开始，参与设计了更大规模的诸多建设项目。第一次进入中国，DCM 设计澳大利亚驻中国大使馆时，设计采用了类型学方法，运用了一些晚期现代主义和后现代主义的元素，这与 70 年代至 80 年代事务所的其它作品相一致。第二次进入中国时，DCM 直接参与了中国地产市场的开发，设计了大量的大型住宅项目，而此工作还在继续。这些项目采用了一些晚期现代主义的形式，至上主义和富有动态的构图，并且使用了类型学方法来营造住宅小区氛围。这些作品属于第三阶段，设计于 2000 年后，但一些早期的构思，特别是类型学方法，再次得到运用并产生了积极的效果。与此同时，在整个 00 年代，DCM 自己的设计语言已经演变到"全球主义"的状态：一种开放、多元、包容众多差异的框架已经显现；一个新的系谱已经产生，其中有一些作品在系谱的两个极端，但同时也与事务所其他作品相互关联；一个关于全球的与地缘的想象也在遍布各国各地的建筑作品中逐步趋于明显；同时，设计思想和交流也更加依赖数码科技手段，使设计文件能飞越地缘距离在全球快速传递。

为更深入研究 DCM 在中国的实践，我们必须从历史角度全面观察事务所的工作，这样才能确定中国项目的关联性，并进一步考察事务所在中国与各方的互动合作情况，以及这种互动合作在更大范围里的作用与影响。

2 "类型学" 阶段（Typology, 20 世纪 70 年代至 80 年代）

虽然 DCM 的作品在 70 年代和 80 年代属于后现代风格，它强调参考历史和侧重传统，但仍然偏向理性或类型学的方向（如罗西（Aldo Rossi）和克列尔（Leon Krier）），而不是装饰性和象征性的方向（如文丘里（Robert Venturi）和葛瑞夫（Michael Graves））。"理性"后现代主义重视都市，并倾向于将建筑物处理成抽象的形式。这包括类型学方法的两个方面：去重构都市空间的类型（广场、公园、大道和小巷），以及采用传承下来的类型去控制建筑物，配合并强调轴线布局的都市空间领域。因此，在这个阶段的 DCM 作品中，都市方面的设计集中于院落、广场或柱廊等特定形成，并控制着建筑物的轮廓。至于单体建筑的设计则趋向抽象化，运用点、网格和方块形体，并配以片墙（开有洞口）和立方体块。更高程度的抽象在往后的发展中继续呈现。整体来说，作品是理性的，关注都市的格局，在 80 年代逐步加强抽象的处理——这一般被视为"晚期现代"的方向。突出的例子都在墨尔本，包括在皇子广场（1985 年）和州立图书馆暨博物馆设计竞赛（1986 年）中采用类型学方式去控制建筑物和都市空间，以及在展览大街 222 号（1985 年 / 1988 年）和科林斯大街 101 号的办公室塔楼（1987 年 / 1991 年）[2] 中采用抽象方法所塑造的现代的作品，尽管其中的裙楼采用了古典或后现代的风格。[3]

就在这个期间，1982 年 DCM 设计了位于北京的澳大利亚驻中国大使馆，该建筑于 1992 年落成。[4] 北京大使馆明显地运用了类型学方法来重建以低矮胡同建筑为肌理的城市形态，它采用了围合的院落布局与老北京的宫殿和四合院建筑传统对话呼应。设计运用了轴线（一个应用在欧洲都市广场和中国宫殿和四合院中的通用的设计手法）、对称、片墙、墙上的方点网格以及墙上的方形大窗洞，使建筑在视线上对外通透，以象征澳大利亚的开放性思想。通过类型学方法，该建筑与中国首都随处可见的传统的围合院落类型进行了交流对话，而这种形式语言又是现代的和抽象的，代表了西方的一种设计话语，同时也表现了建筑所代表的国家的特征。位于东京的澳大利亚大使馆是同时期的另一个作品（1986 年 / 1990 年）。这两座兴建于 80 年代、分别位于北京和东京的大使馆在本土施工技术上的差异，已经多次被专家谈及。[5] 然而，尽管存在差异，这两座建筑在设计上却有相似的空间布局和形式定位——它们都是晚期现代主义作品，都采用了类型学上的轴线和围合，都充分考虑了都市语境和历史文脉。这些相似性并不奇怪，因为它们出自于事务所同一个历史时期。

3 "至上主义"阶段（Suprematism，20 世纪 90 年代）

如果说第一个阶段是后现代并带有类型学和晚期现代主义的趋势的话，那么 90 年代则是一个非常不同的阶段，

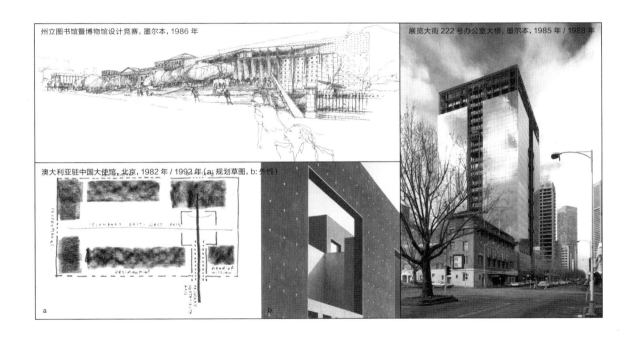

州立图书馆暨博物馆设计竞赛, 墨尔本, 1986 年

展览大街 222 号办公大楼, 墨尔本, 1985 年 / 1988 年

澳大利亚驻中国大使馆, 北京, 1982 年 / 1992 年 (a: 规划草图, b: 外墙)

运用了动态而富有诗意的形式语言,并在 2000 年前得到了逐步的发展和体现。这个改变开始于 80 年代后期。之后的设计,逐步挣脱了历史的枷锁,并开启了对纯粹形式领域的探索。动态的线条、金属的表面以及纯粹的体块已在小型作品中出现,如文森特设计工作室(Emery Vincent Design Studio, 1986 年 / 1987 年);至于框架和立方形式的现代主义的处理,则出现在展览大街 222 号和科林斯大街 101 号的办公楼。然而,决定性的转折,是一件 1986 年设计并完成的桌上艺术装置(这件作品在墨尔本的维多利亚州国家美术馆的艺术装置设计竞赛中获得第一名)。[6] 作为一件艺术品,它是一个对纯粹形式和形式关系的研究,脱离建筑在尺度、功能和施工上的实际考虑。在此,DCM 以"至上主义"为基础,对新的形式语言进行了探索,运用了一些跃动的形态去表达进步、现代性和工业化的观念。这些语言在 20 世纪 10 年代的俄罗斯首先由马列维奇(Kasimir Malevich)和其他艺术家进行探索,并且由先锋艺术家塔特林(Vladimir Tatlin)和利西茨基(El Lissitzky)在 20 年代的构成主义中继续进行探索发展。DCM 的探索,当然是对当时库哈斯(Rem Koolhaas)和哈迪德(Zaha Hadid)对正在兴起的所谓"解构主义"动态形式的兴趣的回应。但是,DCM 能在它们的作品中塑造出属于自己的特点,即对材料和建构更加关注。1986 年的桌上艺术装置运用了一系列银色铝合金的枝条和板片构成一个动态组合,其中有三条长度一致的三角形管状物作水平向伸延,上下随意配以枝条。三角形管状物的表面有点状网格。一座高耸的塔楼在水平向的三角形管状物上拔地而起,周围再配以飞跃的枝条和板片,由此构成一个动态的组合,仿佛一部正在移动的现代机器。

正如建筑评论家贝克(Haig Beck)在 2000 年所指出,这件非建筑的雕塑作品,基本上对 DCM 从 20 世纪 80 年代后期开始的整个设计方式作出了定义。[7] 事务所的建筑作品中,除了垂直和水平构图上的枝条、板片和网格外,还有三维框架、立方体块、随意元素、中介空间和一些标示各部分或元素的鲜艳色彩。这些方面随后被组织成贝克所谓的"元素的组装"(assemble of elements)。建筑因此被解体和重构,借以达到对建筑的抽象,同时具有"至上主义"精神,如 1986 年的装置作品所反映出来的那样。[8]

这个阶段最佳的例子包括墨尔本博物馆(1994 年 / 2000 年)、墨尔本展览中心(1993 年 / 1996 年)和墨尔本市城标(1994 年 / 1999 年)。[9] 前两个作品是大型公共建筑,其中第一个建筑明确运用框架,许多局部体块在其间穿插走动。第三个案例,墨尔本市城标,是一组放在通往城市中心的高速公路两旁的抽象雕塑组合,有一排红色的斜柱(20m 高)、一条黄色的长柱(仿佛吊杆闸口)和一条在公路旁的蜿蜒的黄色的隔音墙。当时速达到 100km/h 时,

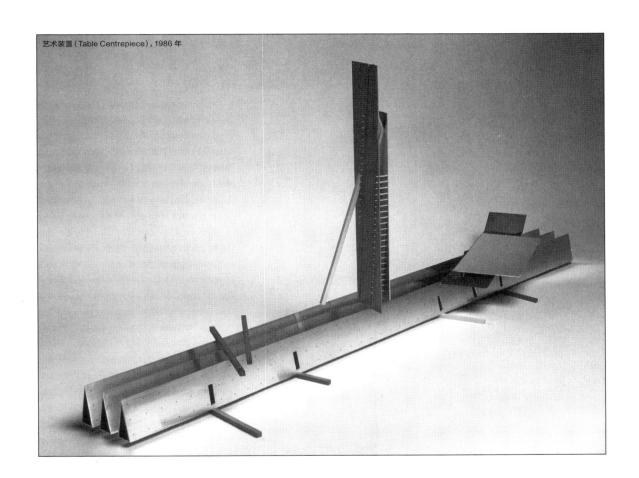

艺术装置(Table Centrepiece), 1986 年

这组雕塑为驾车者带来了精彩的视觉体验。这些新的形式语言在 1986 年之后出现，受到"至上主义"的启发，但通过一系列在澳大利亚的项目得以发展；它们比之前的作品更加统一、更加自信，构成 DCM 一套独特的形式语言。有趣的是，事务所在这个时期并没有在中国设计任何建筑；直到新千禧年的到来，我们才看到 DCM 重新进入中国，而这时与 80 年代的语境和情况已经截然不同了。

4 "全球主义"阶段（Globalism，21 世纪 00 年代）

在 20 世纪 90 年代后期，DCM 工作中新的压力和对速度的要求已经开始出现。到了 21 世纪 00 年代，DCM 进入了一个非常不同的工作环境。当事务所在 2000 年接受邀请进入中国市场后，对设计数量和速度的要求不断提高。正如萨克教授（Leon van Schaik）所指出，DCM 同时在不同国家，包括英国、中国和新加坡等地参与项目，由此出现了一个更宽的"具有不同设计速度"（design speeds）的系谱。[10] 除了时间和数量的压力外，DCM 还面对像中国这样的新兴市场不断创新、持续提供新思路的压力。同时，21 世纪的设计以及文件管理和交流沟通等工作，都增加了对通讯和数码技术的依赖，以求不断在第一时间跨越地理距离，传递信息。这种环境催生了对时间、空间和大地景观的新体验，以及基于世界的对自我和国家的新认识。

结果，在 21 世纪的第一个十年，我们见证了更具极端倾向，在形式语言领域上更加宽广的一系列新作品。在此系谱中，一个极端是更加强调了基本的立方体和笛卡尔的几何正交；而另一个极端则是更加潇洒浪漫的举动，包括三维曲线的游走（如起舞的飞龙）、垂直部件的竖向组合以及面对辽阔大地所采用的超常的水平线条表达对大地的体验。作品的范域，在永恒的静止与加速的前进运动之间。DCM 作品的新，就表现在这个领域或系谱的出现，突出反映在两个极端趋势的出现和许多新元素的运用上。当然，早期的思路手法在此已经被运用和发展了，然而新的突破更加重要，需要明确提出。经过仔细考察，我们可以列举出 DCM 在 21 世纪初在设计上的新突破：作为压力和方法的"量"、部件的垂直组合、龙飞凤舞的曲线、辽阔超长的水平线、柏拉图式立方块体、全球视野中的澳大利亚以及为大规模群体设计的建筑类型（中国项目）。

4.1 作为压力和方法的 "量"

当 DCM 受房地产开发公司"阳光 100"之邀来到中国时，事务所很快意识到设计任务之繁重，因为项目包括了

墨尔本博物馆，墨尔本，1994年/2000年

墨尔本城标，墨尔本，1994年/1999年

大量的住宅单元和庞大的建筑面积；项目的尺度是巨大的，工作的进程又是快速的，许多项目几乎在瞬间进入施工。举个例子，从 2000 年至 2005 年，DCM 在中国各地设计和建成了 25000 个住宅单元，也就是说，事务所每月完成 417 个单元，或每年 5000 个单元。[11] 按照业主的策略要求，这些住宅项目除少数在北京外，遍布全国，但主要集中在二线城市或内陆省份。[12] 其中相对较小的项目，也大约有 1000 个住宅单元，总建筑面积大约 20 万 m^2（包括欧景城市广场和上东国际新城，两个项目都位于南宁，分别落成于 2007 年和 2011 年，前者提供 644 个住宅单元，面积为 21 万 m^2，后者提供 1220 个住宅单元，面积为 18.6 万 m^2）。[13] 而相对比较大的开发项目，总建筑面积可以超过 100 万 m^2，例如重庆国际新城和位于无锡的项目，其总建筑面积分别是 172 万 m^2 和 120 万 m^2（两个项目的第一期分别在 2007 年和 2011 年落成，分别提供 4000 个和 1392 个住宅单元）。[14]

  DCM 如何完成这些规模如此之大、离墨尔本如此遥远的项目？他们如何应对这些无论在政治上、文化上还是语言上都与澳大利亚的环境截然不同的项目？采用一个框架式系统的方式无疑是必然之举，如此才能使本土的力量参与其中。在彼此互动与协商的基础上，设计得以向前推进，以完成大规模的建筑任务，同时还有更多的项目尚在推进中。对此，丹顿（John Denton）视之为"草图"式的运作。[15] 但仔细观察后，我们发现这是一个非常复杂的工作模式：事务所首先提供基本设计框架作为合作探讨的基础，之后由设计院提供本土技术支持作具体落实。期间双方保持不间断的沟通和现场视察，其中也包括了双语职员作出的重要贡献，如龚耕所起的关键作用。"草图"模式提供了一个框架，容纳了"不确定性"，主要是对本土各种元素的考量和当地的技术支持；它也允许既定模式下的重复，以适应住宅单元的巨大规模，但同时这个过程又是高度可控的，在形式语言上具有 DCM 作品的特征，同时又考虑到了中国的"量"的实际状态。我们或许可以视之为一种"快速形态"（fast form）或是一种"简图形态"（diagrammatic form），它允许规模与速度的发生，同时又保持了设计的意义。

4.2 垂直的组合

  建筑评论家们指出，DCM 的作品包括了一种"元素的组装"（贝克）或"部件的交响"（罗伊（Peter Rowe））；这是他们针对 20 世纪 90 年代作品的分析所得。[16] 倘如仔细观察这些设计，我们会发现这些过去的组装基本上是水平向的。所谓交响的结合或组装是沿横向平面展开的，而部件或元素分散在大地表面上。落成于 90 年代后期的墨尔本博物馆、墨尔本展览中心和墨尔本城标是最佳的例子。而 21 世纪 00 年代出现的，是新型的组装，主要特点是其垂

南岸国际新城,重庆,2003 年 / 2007 年

直性，它在高空中在垂直塔楼上发生。贝克曾正确指出组装中包括了解构与重构。[17] 在 2000 年前，它只在水平方向进行，而现在却获得了垂直方向的表达。

其中的一种形式是将塔楼拆成数个细直筒，并将一束修长的直筒组合为一幢塔楼，通常给单个筒井以不同的高度，由此拉长垂直的线条并强化高耸的效果。这在香港的大业大厦（2000 年 / 2002 年）、重庆南岸国际新城（2003 年 / 2007 年）和新加坡的亚洲广场（2008 年 / 2012 年）中最为明显。晶簇综合塔设计竞赛方案（迪拜，2006 年）亦采用这种设计手法。公司早期的作品展览大街 222 号中就已存在这个构思的雏型：它将一座塔楼分为四座，改变部件的比例，令塔楼的比例更加细长。

第二种形式更加戏剧性。这里涉及的部件不是垂直式细长的塔楼，而是垂直的互相平行的几个平面板块，中间夹着不同长度的横向筒管，令整个作品呈现出一个将不同部件紧扎在一起的组合体，而横向筒管的两端向外作不同尺度的伸延，形成凌空飞翔的盒子，介乎于垂直平板之间（这些垂直平板里面是围封的房间或上空的中庭），如曼彻斯特民事法院和北京的中坤大厦，两者都在 2002 年设计，2007 年落成，而前者远比后者精致，且在形式上和材质上更加丰富而有变化。曼彻斯特民事法院可以被认为是 DCM 近年来最好的作品之一，其中值得赞赏的地方包括在空间布局上的经济考虑，垂直体积与横向伸延的创意性构图，对材质和透明性的巧妙精炼的使用以及落成后建筑所产生的整体形象：这一切都给人带来前所未有的惊喜。这座建筑一共获得超过 25 个奖项，包括 2008 年英国皇家建筑师学会（RIBA）颁发的"国家建筑奖"（National Award for Architecture）。

4.3 龙飞凤舞的曲线（弯曲与具象写实的线条）

2000 年后 DCM 的作品明显地运用了更多曲线、曲面和有动势的形态。它们运动的"速度"各有不同。有一些是缓慢而柔和的曲线，如维多利亚花园（海浪公寓）和乐天银泰百货（2004 年 / 2008 年），沿街立面呈波浪形。这两个作品都在北京，分别在 2003 年和 2008 年落成。乐天银泰的玻璃外墙呈波浪形的轮廓，彷佛是舞台上的帷幕，与原址上的京剧戏院呼应对话，亦在当下的戏剧性的都市日常生活语境中，为购物中心的内外，提供了一个视觉沟通的中介。

此外我们还可以看到比较快速的、更有动感的曲线，在三维空间中如龙飞凤舞的游走。这可以在两个设计竞赛方案中看到：长沙的滨江文化公园（2004 年）和杭州的金沙湖步行景观桥（2010 年）。滨江文化公园由一个筒管状的空

亚洲广场，新加坡，2008年/2012年

民事法院，曼彻斯特，2002年/2007年

a　　　　　　　　　　　　　　b

间组成，在河畔舞动和盘旋，后面设有一座垂直的筒状塔楼。金沙湖步行景观桥由三个互相交错的曲线组成，包括两个曲线桥（分别用于运动观赏和通行的功能）和一个支撑结构（与奥雅纳合作设计）。

另一个这种类型的作品是一个三维的运动曲线，但在此却表现了一组花瓣的轮廓——南宁的城标（2000年/2002年），由一组象征花瓣的十片红色金属穿孔板块组成，高度介乎10m至20m之间，散落在公路的一旁，前后相隔的距离超过600m。对于在公路上驱车前往南宁的人而言，它们看似一朵渐渐绽放的莲花；它在人们每次驱车通过时"绽开"。

以上的三个案例都是在21世纪00年代后出现的。在20世纪90年代，曲线大部分是横向的，也不具有很强的动感，如墨尔本城标中沿公路缓缓游动的黄色长墙，以及布里斯班南岸的大型敞廊（1997年/1999年）中小径与网筒的横向蜿蜒。在海浪公寓和乐天银泰百货，水平向的曲线垂直延伸，形成一个大体量的弧形表面。在长沙和杭州的两个竞赛方案中，曲线提升到一个新高度，呈现出大胆的三维表达。至于南宁市城标的红色板块，由于外形与莲花相似，所以是一个具象的描写，这是DCM形式语言中一个新的时刻：DCM的作品一般都是抽象的，而这个形态处于抽象的边缘。有趣的是，每朵花瓣都是由两块红色钢板构成的，它们是抽象的，也是具象的，在两者之间，同时又集结构和装饰于一身。[18] 它们处于两个领域的边缘，位于建筑形式语言的最前线，在模糊两个领域（抽象与具象，结构与装饰）中，探索对情感和生命的表达。从这个角度来看，这个类别的所有案例，包括波动的帷幕和龙飞凤舞的曲线以及早前的实验，是同一母题在不同阶段的发展，它是一种对生命形式的新表达。

4.4 超长的水平线

在作品中运用绝对水平或微微弯曲的线条，分别反映在墨尔本的香克海角别墅（1997年/1999年）和联邦广场竞赛方案（1997年）上。作为一个概念，1986年的艺术装置是运用超长横向线条的早期作品。然而，这个雕塑的戏剧性或强度在2000年之前的建筑上尚未得到体现。之后我们所见证的，是一个更加明确的横向长线的表达，它可以是绝对的水平线，也可以是广阔的延绵起伏的缓坡的一部分。横向长线条不但展示了笛卡尔式或柏拉图式的纯粹，同时更进一步地体现了一种速度感和永恒性。可以说，这些超长的横向水平线，表现的不仅是我们眼前所见的辽阔风貌，也是一个关于时间、空间和大地的21世纪的新认识；今天，我们彼此生活在同步的网络的各个点上，互相间在全球范围内永远而即时地联系在一起。

滨江文化公园设计竞赛,长沙,2004年

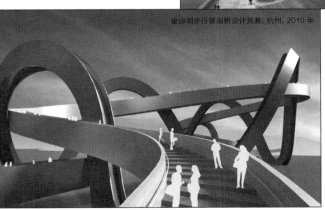
金沙湖步行景观桥设计竞赛,杭州,2010年

乐天银泰百货,北京,2004年/2008年

落樱酒庄盛酒店，北京（延庆），2010 年

梅德赫斯酒庄别墅，墨尔本（亚拉河谷），2002 年 / 2008 年

丽山别墅与酒庄，墨尔本（亚拉河谷），2010 年 / 2012 年

关于绝对的水平线，我们可以列举位于墨尔本的梅德赫斯酒庄别墅（2002年/2008年）和丽山别墅与酒庄（2010年/2012年）。两者都采用了悬挑结构创造了一个平台或方筒。两者在水平伸展和基本定位上都是自觉的"绝对主义者"。比如，丽山别墅与酒庄的上层方筒和下层基座分别采用了精确的东西向和南北向，形成一个笛卡尔的几何正交。悬挑结构在前后分别向外伸出6m和9m，而在梅德赫斯酒庄别墅则向外伸延达11m之多，这是21世纪00年代所完成的一个惊人之举（都运用了金属板块）。

说到修长而微弯的"缓坡"，这方面例子有澳大利亚战争纪念馆（堪培拉，1999年/2001年），英国巨石阵游客接待中心暨博物馆（第一方案，英国，2001年）以及落樱酒庄（方案设计，北京，2010年）。这些作品传递出一种对辽阔地貌的认识和一种我们在信息技术网络上飞越大地与海洋的地理阻隔的速度感。

4.5 柏拉图式的立方体：

在立方块体的运用上，我们也能在90年代或早期的一些作品中发现其雏型的表达，比如文森特设计工作室和史华慈画廊（Anna Schwartz Gallery，墨尔本，1993年）。两者都用了金属或混凝土方块体作为室内的元素；另一个例子是墨尔本博物馆，使用了比较轻快的仿佛在摆动的大型立方体。21世纪00年代所呈现出来的，是在尺度和姿态上对方块体或方型构造更加明确的使用；其姿态明显是柏拉图式或笛卡尔式的，强调重量感、绝对性和永恒性，似乎要再次把我们拉向土地，并与时间、速度和运动的轻快的新感觉构成强烈的对比。无论如何，柏拉图式的立方体的运用明显地增加了，这些案例有：布里斯班广场（2001年/2006年）、中国国家博物馆（竞赛方案，北京，2004年）、澳大利亚驻雅加达大使馆（2009年/2015年（预计））、新加坡科技设计大学（竞赛方案，2010年）以及威尼斯双年展的澳大利亚馆（2011年/2015年（预计））。从这十年对立方体应用的频繁程度来判断，可以肯定地说，它在事务所的作品系谱中扮演着一个重要的角色；这个系谱是出现在21世纪00年代的一个新范围，这个范围从柏拉图式的沉重一直延伸到速度与运动的轻快。

4.6 全球视野中的澳大利亚：

在DCM逐年增加的海外设计发展过程中，一种在全球视野下对本土的理解更清晰了。事务所21世纪00年代的作品呈现出了关于澳大利亚身分认同之构成，以及如何在建筑上精确表达这种对本土认同的明确理解。这个国家的特征包括辽阔的大地和丰富的矿产，可以在建筑上分别表现在无尺度的形态和形态上的金属质地。这个形态通常是柏

中国国家博物馆设计竞赛，北京，2004 年

新加坡科技设计大学设计竞赛，2010 年

拉图式的方体或方块盒子。如果说去除了细部处理的规则方体或方块盒子代表了天地的辽阔无垠的话，那么，金属质地的方盒子则象征着矿产之丰富。显然，最典型的例子是威尼斯双年展的澳大利亚馆：它是一个大胆又富有纪念性的方体，自律而无尺度，隐喻一个辽阔的国土，建筑实际的也是象征意义上的故乡。这个方盒子的表面由南澳的黑色花岗石覆盖，赋予了澳大利亚展馆神秘、古朴和宏大之感，具有多层的涵义，有待参观者去发掘。另一个例子是澳大利亚驻雅加达大使馆。这个大使馆有五个方块体，每个方块体用不同金属材质所覆盖——锌、铜、黄铜、钢和铝，展现了国土的富饶与广阔。

4.7 服务百万人的建筑类型（中国项目）

　　DCM 在 21 世纪 00 年代的作品中有两个处理大尺度或大规模的方法：一个是无尺度的单体形态，表达隐含的海量特征和开放性；另一个是具有实质数量上的形态，提供数以万计的空间单元。后者实际上就是一个提供给数以万计的住宅单元、有百万平方米建筑面积的类型手段。这种反映真实数量的类型与形态出现在事务所在中国的建筑作品中。"草图式"的框架式的方式容纳了当地的变数，以及当地设计院的贡献；总体设计的框架是开放的，有一定的宽容性，同时在总体效果上严格地受墨尔本事务所总部的把控。DCM 的一个重要设计手段是经过精心设计的有序复用，如此塑造了一个有韵律的且强有力的城市形象，并创造了有效地提供市场需要的规模。其结果就是在中国十多个城市中完成了多达三万五千个单元、拥有数百万平方米的大型居住建筑。DCM 在此作出了两个主要的贡献。首先是为这些"量"提供了一种形态、秩序和标志，使之具有严谨的形式章法：包括"至上主义"的线条和板片，标示不同建筑或建筑部分的鲜明色彩，以及配合空间布局的对体量的形态控制，经心控制的有效重复以创造空间序列。这些手法的一贯使用，使这一系列的住宅开发项目获得了一个明确的形式上的性格和标志。"阳光 100"遍布全中国多个城市的住宅建筑，获得了一个特有和出众的形象。

　　DCM 在这些开发建设项目中的第二个贡献，是市区和市郊由数千居住单元集合而成的都市尺度的类型学格局。这亦是比较重要的贡献，因为它涉及到空间的整体性。DCM 的设计中，从来没有放弃，而是一直自觉地将 20 世纪 70 年代至 80 年代的类型学思想用作空间组织方法，以构成特定的住宅区和都市空间；这些区域格局包括许多元素，如围合、入口、中轴线、视觉通廊、台地高程的变化、作为"墙体"的板楼等等基本要素，形成了各种院落或广场（在欧洲的市民空间和中国大大小小的合院体系中都可以发现这种空间手法），以及通过重复创造的空间韵律。[19] 这也是 20

澳大利亚馆,威尼斯双年展,2011年/2015年(预计)

(数码集成照片;a 在澳大利亚;b 在威尼斯)

世纪70年代至80年代后现代主义所留下的最主要的遗产之一，罗西和克列尔曾对此进行探索，它随后在DCM的作品中得以实践和检验。这在澳大利亚驻北京的大使馆中已经出现，随后在21世纪00年代以更大的规模再度出现，更直接地参与到中国的建筑设计中。这种类型学思想有其重要的实用价值，因为它提供了一个设计方法，以抗衡城市肌理中出现的具有摧毁性的破坏现象，即在开敞空间中大量放置无关联塔楼的现代建筑手法。这对当代中国特别有重要意义，因为拆毁的趋势是如此的普遍和迅猛。

在这方面，最佳的案例应该包括重庆南岸国际新城（2003年/2007年）和南宁欧景城市广场（2001年/2007年），这两个项目分别提供4400（建成的）和644个住宅单元，总建筑面积分别达到172万 m^2 和21万 m^2。南宁上东国际新城（2005年/2011年，1220个住宅单元，总楼面总积是18.6万 m^2）亦是一个好例子；其中有一个围合的都市小区，有多个入口导入、内部设小巷、兼有台地高程的变化，小区庭院的布置，以及一排蜿蜒曲折的板楼作为一幅大墙体在后面界定领域的范围。

DCM在中国的作品不局限于住宅小区的建设，然而这些庞大的项目却构成了事务所在当地的主要工作。对比事务所在其他地方尤其是在澳洲和英国的作品，这些在中国的大项目相对而言是比较粗线条的，有时是比较简陋的；这或许是数量大、时间有限、工程预算的局限，以及二线市场的目标定位所致。然而，这些项目的都市类型的建构、形式秩序的组织和视觉标志的树立，无疑都已经清楚地实现了。所用的"草图"的框架式的方法具有包容性，同时在设计上又是高度可控的。这里有一个复杂的设计知识的"传递"过程（'transfer' of design knowledge）；但是，在其本土化的过程中，它又是一个互动和综合的状态，将西方传来的严谨的形式秩序和亚洲市场的动力结合在一起。[20]

5 加速的现代主义和中国项目的作用与意义

就事务所的整体工作而言，因为工作速度的加快和遍及全球的快速通讯工具的到来，加上澳大利亚与亚洲和欧洲更紧密的全球互动，DCM在20世纪90年代的"至上主义"已经发展成21世纪00年代的全球主义构架，并表现为以下三个主要特点：第一，对不同设计速度的全球兼容，包括英国缓慢而精致的作品和中国快速而粗线条的建筑；第二，对澳大利亚及其辽阔地理景观的新理解，包括对金属材料、超长的水平线条和无尺度立方体的使用；第三，一个更宽广的设计语言系谱，从动态（三维运动、垂直组合、为百万人设计的快速形态）到柏拉图式的永恒静止，这个系

城市广场，南宁，2001年 / 2007年

上东国际新城，南宁，2005年 / 2011年

56　加速的现代主义

谱最近十年才出现，尽管它由过去演变生成。产生这些新构思和趋势的一个重要背景正是新时代的速度和压力，而由此看来，对于事务所这十年设计生产的具体工作环境而言，中国应该是一个重要的推动因素。

衡量中国在 DCM 整体工作中的关联性，我们可以从两方面来观察：一般的相关性和直接的参与互动。就一般的相关性而言，可以这样说，中国项目的高速度，对事务所的整体视野产生了潜在的影响或作用，强化了一般意义上的速度感。作为全球发展最前沿的中国，可以说在高效工作、追求急速创新方面表现突出，当然这亦是全球现象的一个趋势。其次，在应对当前压力和市场需求而进行的直接参与这方面，DCM 为此至少作出了四点贡献：一，服务于百万人的住宅类型；二，"龙飞凤舞"的动态造型实验；三，对文化的回归，以及关注传统和地貌的小型而精致的建筑作品的出现；四，通过在中国、英国和澳大利亚同步进行相似的设计构思，来消解中国与西方之间的阻隔。

关于"百万人的建筑类型"与"龙飞凤舞"，前面已经谈及。DCM 在中国的设计工作的第三个贡献是对文化议题的回归。在澳大利亚驻北京大使馆的设计之后，有一些规模较小但却精致的作品同时出现，以表达文化的传统或对辽阔乡野的体验。这里的案例包括在北京中心原吉祥戏院旧址上新建的乐天银泰百货的玻璃幕墙，以及北京郊外的落樱酒庄。落樱酒庄以细腻的设计手法，结合北京西北巍峨山峦之下延绵起伏的地貌，并用超长的多层墙体创造出了田野中富有诗意的体验。

第四个贡献是消解中国与西方之间的差异，通过在中国、英国和澳洲本土同步处理方法相似的构思，将中国的作品放在全球设计话语的平台上。举例而言，北京中坤大厦和曼彻斯特民事法院几乎同时设计同时建成（2002 年／2007 年），两个项目的设计都对大型板块和长方盒子的垂直组合的构思进行了探索。同样地，运用超长水平线条去呼应开阔的大地景观，亦出现在落樱酒庄、丽山别墅与酒庄以及英国巨石阵游客接待中心第一和第二方案（2001 年和 2009 年）之中。在这些案例中，中国的作品通过事务所在不同国家的实践进入了全球设计话语的平台。当然这也是一个大趋势，在中国的建筑设计，无论出自中国建筑师还是其他国家的建筑师（如王澍、张永和、崔恺、都市实践、哈迪德和库哈斯），都大量进入了全球话语之中。DCM 的跨境构思和设计传递，是这个发展的一部分。当然，DCM 的整体贡献包括了四个方面，其中第一方面，即为中国各省市的百万人设计的"快速形态"，包括"新城"和"城市广场"等都市空间类型，可以说是最有特色的。

如果比较事务所在中国、英国和澳洲的作品，尤其以墨尔本和曼彻斯特所完成的精美作品来判断，必须承认，事务所目前在中国的建筑虽然数量上是庞大的，但形式和建构设计上还缺乏更加精致的美学质量。考虑到 DCM 在其他地方的作品质量之精美，我们应该承认这里还有空白需要填补。可以预料，DCM 在中国的设计，会出现更精致的质量和更明晰的理论定位，随着社会的发展，中国自身也愈加关心"创新"。而为了创新，我们应该鼓励更多在设计、话语和创作方面的努力。

  作为一个拥有全球百分之二十人口、经济规模排名第二、发展速度最快的一个国家，中国是正在兴起的具有自身独特"强度"的一个独特的"高地"（借用德勒兹 (Gilles Deleuze) 和瓜塔里 (Félix Guattari) 的用语）；它正在发展一个新的、包容千百万人的现代构架，有自己的尺度、压力和问题，或许还有自己独立的价值观和审美观——这是一个前所未有的的现代体系。中国在建筑设计话语上的贡献目前尚未出现，但会在未来的几十年出现和成型。如何去理解中国并与之打交道，如何建立和协助发展一套有中国参与的设计话语，可以说是当今世界建筑学科中最重要的议题之一。从这个角度讲，DCM 在中国的工作，作为设计知识的传递，也作为澳大利亚（和英美西方世界）传承的形式秩序和中国社会市场发展动力间的互动和综合，是这个时代一段最重要的旅程。

朱剑飞，

墨尔本大学建筑学副教授，博士及硕士研究生导师。

作者衷心感谢墨尔本丹顿·廓克·马修建筑设计事务所的丹顿和龚耕。他们慷慨地与作者分享了有关材料和想法。同时也要感谢"阳光 100"的工作人员，包括南宁的尹筱筠、周湘和黄剑，重庆的王嵘，北京的唐欣宇以及天津的张洁和白金虎；感谢他们为作者在 2011 年 6 月至 7 月期间进行的调研所提供的大力支持。这篇文章是"澳洲——中国都市建筑设计合作"研究项目的一部分，由墨尔本大学研究合作基金资助（2010—2012 年），特此致谢。

本文由周庆华从英文原文译成中文。周庆华，墨尔本大学建筑与规划学院博士研究生，香港注册建筑师。

1 这些书刊包括：Haig Beck and Jackie Cooper (eds) Australian Architects: Denton Corker Marshall – A Critical Analysis, Canberra: Royal Australian Institute of Architects, 1987; Haig Beck/Jackie Cooper, Peter G. Rowe and DeyanSudjic, Rule Playing and the Ratbag Element: Denton Corker Marshall, Basel: Birkh user, 2000; 以及 Leon van Schaik (ed.) Non-Fictional: Denton Corker Marshall, Basel: Birkh user, 2008。另有一本关于 DCM 的中文专著：聂建鑫，陈向清. 澳大利亚 DCM 作品实录. 北京：中国建筑工业出版社，2002。另外也可参考 Denton Corker Marshall, expresso< expressway: Denton Corker Marshall's non-architecture, Melbourne: Denton Corker Marshall, 2006。

2 这两个时间，"1985 年 / 1988 年"，分别代表设计和落成的年分。这篇文章的其他地方都是按这个方式标示建筑的设计和落成时间，除非另外注明。

3 参考：Beck and Cooker, Australian, pp. 23-6, 31-6, 37-42 and 51-4；四个片断分别讨论了展览大街 222 号办公楼、科林斯大街 101 号办公楼、州立图书馆及皇子广场。关于两个办公楼，亦可参考 Beck/Cooker et al, Rule Playing, p. 116。

4 参考：Beck/Cooper et al, Rule Playing, pp. 86-7, 90-5。亦可参考 Jianfei Zhu, 'Denton Corker Marshall in China: Interactions', in Leon van Schaik, Non-fictional, pp. 134-40，以及 Jianfei Zhu, 'Export or Dialogue', Architecture Australia, vol. 99, no. 5, Sept-Oct 2010, 97-8。

5 参考：丹顿，"访谈"，墨尔本，2011 年 11 月 9 日；亦可参考 Beck/Cooper, Rule Playing, pp. 88-89, 96-107。

6 参考：Haig Beck and Jackie Cooper, 'Development of a Formal Language', in Beck/Cooper, Rule Playing, pp. 33-7。

7 参考：Beck and Cooper, 'Development', in Beck/Cooper, Rule Playing, p. 33。

8 参考：Beck and Cooper, 'Development', in Beck/Cooper, Rule Playing, p. 34。

9 参考：Beck/Cooper, Rule Playing, pp. 40-63, 188-9, 194-209, 223 and 232-37。

10 参考：Leon van Schaik, 'A Tale of Twined Cities', in Schaik, Non-Fictional, pp. 8-23。

11 材料由 DCM 事务所于 2005 年提供。

12 以 2011 年 6-7 月我与南宁、重庆、北京和天津"阳光 100"各位管理人员的访谈为依据。

13 材料由 DCM 事务所于 2007 和 2011 年提供。

14 材料由 DCM 事务所于 2007 和 2011 年提供。

15 丹顿，"访谈"，墨尔本，2011 年 11 月 9 日。

16 参考：Beck and Cooper, 'Development', in Beck/Cooper, Rule Playing, pp. 34, 34-7；以及 Peter G. Rowe, 'Rule Playing and the Ratbag Element', in Beck/Cooper, Rule Playing, pp. 19, 13-32。

17 参考：Beck and Cooper, 'Development', in Beck/Cooper, Rule Playing, p. 34。

18 有关城市大门或城标项目中"写实"问题的讨论，请参 Zhu, 'Denton', in Schaik, Non-fictional, pp. 134-40 以及 Zhu, 'Export', Architecture Australia, pp. 97-8。

19 有关北京澳大利亚大使馆设计的讨论，请参考 Zhu, 'Denton', in Schaik, Non-fictional, pp. 134-40 以及 Zhu, 'Export', Architecture Australia, pp. 97-8。

20 丹顿在访谈中，多次提及设计过程中技术与知识的"传递"（丹顿，"访谈"，墨尔本，2011 年 11 月 9 日）。

**MODERNISM ACCELERATED:
DENTON CORKER MARSHALL
IN THE 2000s
AND THE RELEVANCE OF CHINA
AS A SITE AND A CATALYST**

Denton Corker Marshall is one of the most prominent Australian architectural firms practising today. Founded in 1972, with the main office in Melbourne and branches in London and Jakarta, the practice has produced critically acclaimed buildings both in Australia and internationally, in the UK, China, Europe and Southeast Asia. Although this is well known, and well documented in the publications of the late 1980s, and in 2000 and 2008,[1] the practice is producing new buildings, with new surprises, at a rapid rate. With a limited historical distance, these recent studies have yet to fully capture the practice's development of the 2000s with a comprehensive and analytical perspective. With the benefit of hindsight today in 2013, it is clearer that, for the practice, the 2000s is not merely a continuation of the 1990s, but a new phase with new ideas and perspectives that have emerged.

For example, in the early 2000s, Denton Corker Marshall became involved with large-scale projects in China for the first time, designing housing developments that were constructed in more than ten cities across the country. The Chinese projects were a sharp contrast with those in the UK being designed and built at the same time. In addition, a new sensitivity for landscape and of the world also emerges in the global practice, generating a new understanding of the nation, as evidenced not only in the houses in rural Victoria but also in the Jakarta Embassy and the Australian Pavilion in Venice (both in construction). A new study on Denton Corker Marshall should address these recent breakthroughs, the practice's emerging sense of space, land and geography for a new global perspective, and the practice's involvement with other countries – especially China as a radical or extreme case in the whole spectrum of practice.

What has happened, and how has Denton Corker Marshall evolved in the 2000s? What marks the practice's key contribution beyond a mere extension of the 1990s? What position does China occupy as a site and a market for the practice's overall portfolio? What has been produced in China? What is brought to China and what is required of the practice for the work there to succeed? How did the market and the designer interact with each other? What are Denton Corker Marshall's contributions, and what is China's role as a catalyst, if any? This essay aims to address these questions, in an attempt to understand both Denton Corker Marshall's overall work, as well as China, as a site and a catalyst, in the first decade of the 21st century.

An Outline

With a hindsight available today, it is clear that Denton Corker Marshall's development includes three phases: a phase of 'typology' in the 1970s-1980s, in a post-modern framework with late-modern ideas (abstract planes and forms); a phase of 'suprematism' in the 1990s, along with de-constructivist ideas occurring elsewhere; and a new phase of 'globalism' in the 2000s, with an acceleration characteristic of the new century. The divisions are not clear-cut, and many ideas flow into the new frameworks later on – there is an accumulation of ideas as well as a repudiation of some previous elements. On the whole, though, the late-1980s and early 1990s mark a decisive departure from the work produced earlier, with the formation of a language unique to Denton Corker Marshall; whereas the 2000s are not merely a continuation but rather a significant ascendance, with new pressures and demands for speed absorbed, and new experiments and ideas developed. I would like to call the third phase 'globalist' – not because there was no international work before, but because in the recent period there is a greater assertion of Melbourne-based ideas around the world with more overseas involvement in practice, a parallel rise of a heightened sense of the world and the geographical, and a use of faster telecommunications in work for the practice and in daily life all around us.

In this three-phased development, China comes into the picture in 1982 and after 2000, that is, the office has entered China twice, in 1982-1992 for the Australian Embassy, and then, at a much greater scale for many projects, from 2000 onwards. In the first encounter, the office designed the Australian Embassy in Beijing, where it adopted a typological approach with late-modern and post-modern elements which were consistent with the practice's other works of the 1970s-1980s phase. In the second encounter, which is still ongoing today, the interaction is directly involved with the market in China, on a much larger scale, with many housing development projects. These projects adopt late-modern forms, suprematist or dynamic compositions, and a typological approach to the formation of residential communities. The work here is taking place in, and is a part of, the third phase of the 2000s, although earlier ideas, that of typology especially, were employed to a constructive effect. At the same time, throughout the 2000s, Denton Corker Marshall's own design language has evolved into a new 'globalist' position: an open and diverse framework has emerged, which is tolerant of great differences. This new spectrum has resulted with works at extreme ends of this position comfortably coexisting in the same portfolio; a global and geographical imagination is increasingly apparent in the works as they are distributed across continents; and, of course, there is more reliance on digital technologies for documenting and communicating design work as it is swiftly transmitted across geographical distances around the world.

To study Denton Corker Marshall's work in China more closely, we must observe the office's portfolio historically as a whole, before we can ascertain how the Chinese case is connected and inquire further on the interaction in China and its implication on the wider picture.

Typology (1970s-80s)

Although Denton Corker Marshall's work during the 1970s and 1980s was post-modern in its emphasis on historical reference and gravity towards tradition, it was a post-modernism on the rational or typological side (as in Aldo Rossi and Leon Krier) rather than the decorative and metaphorical end of the spectrum (as in Robert Venturi and Michael Graves). This 'rational' post-modernism focused on the urban and tended to treat buildings as abstract forms. It included a typological approach which aimed to reconstruct urban spatial types (plazas, parks, avenues and laneways), and to control buildings with an inherited type to conform to a defined urban realm articulated with axes. As such, in Denton Corker Marshall's work of this period, the urban aspect of the design focused on a defined formation of a court, plaza or promenade with the outline of the buildings controlled, whereas the design of individual buildings tended to be abstract in the use of dots, grids, squares, planes (with cut-outs) and cubic volumes, with a higher degree of abstraction emerging later on. On the whole, the work was rational and urban-oriented, with an increasing abstraction into the 1980s – a direction that is normally regarded as 'late-modern'. The prominent examples, all in Melbourne, include Princes Plaza (1985) and State Library with Museum competitions (1986) for a typological approach to control both buildings and urban spaces, and the office towers at 222 Exhibition Street (1985/1988) and 101 Collins Street (1987/1991)[2] for the use of abstraction as a modern work even though the podiums were classical or post-modern.[3]

It is during this time that Denton Corker Marshall designed the Australian Embassy in Beijing in 1982, completing its construction in 1992.[4] The Beijing embassy is clearly a work which employs the idea of typology for a reconstruction of the urban profile of low-rise hutongs and a walled compound to echo the courtyard houses and the palaces of old Beijing. Axes were employed (a universal instrument for both urban plazas in Europe and enclosed courts and houses in China), along with symmetry, planar walls, a grid of dots on the wall, and large square cut-outs in the wall, allowing views into the compound and representing the Australian idea of openness. Using a typological approach, the work was able to communicate with an ancient Chinese tradition of walled compounds found in the Chinese capital, in a language that was contemporary and abstract, which represented a western design discourse – and also the identity of the country it represented. The Australian Embassy in Tokyo was another project of the same period (1986/1990). Much has been said on the differences in local technology and building skills in Beijing and Tokyo in the 1980s.[5] However, despite the difference, the buildings shared a similar approach to spatial planning and formal orientation – both are late-modern works with a typological use of axes and enclosures and a consideration of urban context and history. The similarities are not surprising given the period in which they were designed in the office's history.

Suprematism (1990s)

If the first phase was post-modern with typological and late-modern tendencies, the 1990s was a very different phase, with a dynamic poetic language gradually developing and materialising by 2000.

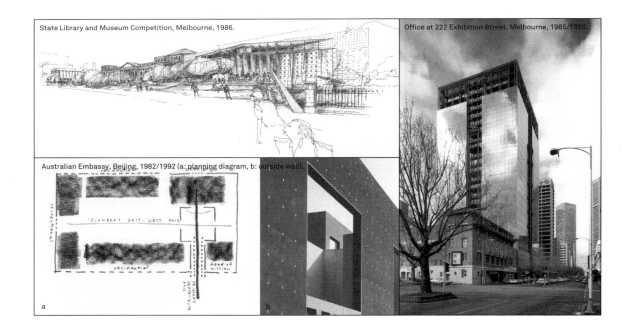

State Library and Museum Competition, Melbourne, 1986.

Office at 222 Exhibition Street, Melbourne, 1985/1988.

Australian Embassy, Beijing, 1982/1992 (a: planning diagram, b: outside wall).

The change began in the late 1980s. From then on, there was an unshackling of the weight of history, and a launch of new forays into the realm of pure forms. Dynamic lines, metallic surfaces and purer volumes emerging in smaller works such as Emery Vincent Design Studio (1986/1987), while a modernist treatment of frames and cubic forms is found in the office towers at 222 Exhibition Street (1985/1988) and 101 Collins Street (1987/1991). The decisive break, however, is a table centrepiece designed and constructed in 1986 (which won the first prize in a competition at the National Gallery of Victoria in Melbourne).[6] As a work of art, it is a study of pure form and formal relations, freed from the practical concerns of scale, function and building construction. Here Denton Corker Marshall explores a new formal language based on Suprematism, which uses dynamic forms to express progress, modernity and industrialisation, as first explored in Russia in the 1910s by Kasimir Malevich and other artists, and Constructivism of the 1920s by pioneers such as Vladimir Tatlin and El Lissitzky. This was surely a response to the rising interest in dynamic forms labelled as 'de-constructivism' found in the works of Rem Koolhaas and Zaha Hadid around the same time. But Denton Corker Marshall put its own stamp on the project – it was more materially based and more tectonic. The Centrepiece (1986) was a series of sticks and blades, made of silver and aluminium, assembled in a dynamic manner; it had three long and uniform tubular wedges stretching horizontally, with random sticks beneath and above, and a grid of dots on the surface of the wedges, and then a soaring tower rising vertically from the horizontal wedges, with other blades and sticks flying around, forming a dynamic assemblage reminiscent of a modern machine on the move.

As architectural critic Haig Beck pointed out in 2000, this sculpture, this non-architecture design, basically defined Denton Corker Marshall's entire approach from the late 1980s onwards.[7] In the constructed architectural work of the office, apart from sticks, blades and grids in a vertical and horizontal composition, there were also three-dimensional frames, volumetric objects, random elements, spaces in between, and some bright colours to mark the identity of some parts or elements. These aspects were then organised as an 'assemble of elements', according to Haig Beck, in that a

Table Centerpiece, 1986.

building was de-constructed and re-assembled so as to arrive at an abstraction of architecture, with a Suprematist spirit, as reflected in the Centrepiece of 1986.[8]

The best examples of this time include Melbourne Museum (1994/2000), Melbourne Exhibition Centre (1993/1996) and Melbourne Gateway (1995/1999).[9] The first two are large public buildings, with the Melbourne Museum especially strong in expressing the idea of a framework with parts passing through and around. The third, the Melbourne Gateway, is a sculpture on the highway into the city, with a row of red sticks (20 metres high), a yellow stick (reminiscent of a boom gate), and an orange sound wall curving along the highway; together they form a visual experience for motorists travelling at 100 kilometres per hour. This new language emerged after 1986 – inspired by Suprematism, but developed through a range of projects in Australia – was far more coherent and confident than those in the previous works, and can be said to be a systematic language unique to Denton Corker Marshall. Interestingly, the office did not design anything for China in this period – it is in the new millennium that we see Denton Corker Marshall's re-entry into China, in a context and situation that is very different from the 1980s.

Globalism (2000s)

The momentum of speeding up with new pressures was already present for Denton Corker Marshall in the late 1990s. In the 2000s, the practice worked in a very different environment. There was a rising pressure to design large quantities at a fast pace for China after the office accepted an invitation to enter the market in 2000. There was now a greater spectrum of 'design speeds' – as Leon van Schaik has pointed out – the office had to work simultaneously with different countries such as the UK, China and Singapore.[10] There is also pressure to constantly offer new ideas in the emerging markets, especially in China, besides the pressure on time and demand for quantity. In addition, there is also an increasing reliance on telecommunications and digital technologies to design, document and communicate swiftly, around the clock, across huge geographical distances – a situation that generates a new sense

Melbourne Museum, Melbourne, 1994/2000.

Melbourne Gateway, Melbourne, 1994/1999.

of time, space and landscape, and of one's own place and nationhood in the world.

As a result, during the 2000s we witness a new range of work, much broader than before, with extreme tendencies. On the one hand, it has a stronger emphasis on the primary cube and Cartesian intersections, on the other it has a more assertive and daring play of 3-D curves (waving like dancing dragons), vertical assemblages of parts high in the air, and extra-long horizontal lines for a sweeping horizon in relation to the land, understood in a new perspective. The range is between an eternal standing-still and an accelerated movement forward. This range is new, in its extreme tendencies clearly marked and in the new elements employed. For sure, earlier ideas have been employed and developed here, yet the new preoccupations must be acknowledged. At a closer inspection, we may enlist these as the new breakthroughs of the 2000s: quantity as pressure and as method, vertical assemblage of parts, dancing dragons, sweeping horizontal lines, the Platonic cube, Australia in the world, and a typology for the millions (in China).

Quantity as pressure and as method: Once Denton Corker Marshall was invited to design for Chinese developer, Sunshine 100, it quickly became apparent that the pressure of the task was enormous, because of the huge quantity of housing units and floor areas to be covered, the scale of the projects, and the speed required to resolve the design as the projects went into construction almost instantly. Between 2000 and 2005, for example, Denton Corker Marshall had designed and built 25,000 residential units across China, that is, the office was completing 417 homes every month, or 5000 flats or apartments every year.[11] These housing developments are located across China, many in second-tier cities or inland provinces, as the developer strategically desired, besides a few in Beijing.[12] The relatively 'small' projects have about 1000 homes, totalling around 200,000m^2 (such as the Euro-City Plaza (2007) and Shangdong International New Town (2011), both in Nanning, with 644 homes/210,000m^2 and 1220 homes/186,000m^2 respectively).[13] For the larger developments, the total floor area can reach over one million square metres, such as the Chongqing Southbank International New Town (2003/2007), and in Wuxi, with 1,720,000m^2 and 1,200,000m^2 in their total area respectively (the first stage of both were

Southbank International New Town, Chongqing, 2003/2007

completed in 2007 and 2011, with 4400 and 1392 homes constructed).[14]

How did Denton Corker Marshall cope with these huge quantities of work, on a site so remote from Melbourne, with a political, cultural and linguistic environment so different from Australia's? A framework approach, allowing local contributions, is inevitable. After interactions and negotiations, the collaboration develops and the huge buildings are built, with more underway. John Denton refers to this as a 'sketch'.[15] On closer inspection, however, it turns out to be a very complex format of work; the practice presents design ideas as a framework to be negotiated and further implemented with local knowledge by Design Institutes (known as DIs), with ongoing telecommunications and site visits on the way, with bi-lingual staff making crucial contributions in between, such as Greg Gong's indispensible work. The 'sketch' approach, providing a framework, allows for 'uncertainties', mainly local considerations and technical contributions, and a certain degree of repetition to absorb the number of housing units, but is at the same time still controlled, with the whole product recognisable as a work of Denton Corker Marshall, though it also internalises the Chinese condition of quantity. We may regard it as a 'fast form' or a 'diagrammatic form'; it allows scale and speed to occur, while still maintaining a meaningful sense of design.

Vertical assemblage: Critics have already pointed out that Denton Corker Marshall's work includes an 'assembly of elements' (Haig Beck) or an 'orchestration of parts' (Peter Rowe) when they were analyzing the work of the 1990s.[16] When we observe these designs carefully, it is clear that these assemblages were basically horizontal – the orchestration or assemblage occurred across a horizontal plane for the parts or the elements scattered on the land surface. The Museum, the Exhibition Centre and the City Gateway, in Melbourne, completed in the late 1990s, are the perfect examples. What emerges in the 2000s is a new kind of assemblage that is characteristically vertical, occurring high in the sky for soaring towers. Haig Beck was right in saying that there is a de-constructing and a re-constructing in the assemblage.[17] While that was going on horizontally before 2000, it now acquires a vertical articulation.

Asia Square, Singapore, 2008/2011.

Manchester Civil Justice Centre, Manchester, 2002/2007(a,b)

One version of this is to take a tower apart into several slender tubes, and to re-assemble them as a 'bundle' of slender towers, often with shifting heights, thus elongating the vertically lines and intensifying the soaring movement. This is evidenced in Tai Yip Tower in Hong Kong (2000/2002), Chongqing Southbank International New Town (2003/2007) and Asia Square Tower in Singapore (2007/2012). The Cluster Complex Competition (Dubai, 2006) may also be seen as a version of this. In an early work at 222 Exhibition Street, Melbourne (1985/1988), there was an embryonic form of this idea: the separation of one tower block into four, to modify the proportion and to make it look slender.

The second version is more dramatic: the parts here are not vertical tubes (slender towers) but planes and boards layered upright with horizontal tubes of different length stacked in between; the whole work is a tight bundling of these parts, with the ends of the tubes projecting in and out as flying boxes, but are also sandwiched between flat planes and boards which are closed rooms or void atriums within. The examples are Manchester Civil Justice Centre and Zhongkun Tower in Beijing, both designed in 2002 and completed in 2007, although the first is far more refined, with a greater richness in formal and textual variety. Manchester Civil Justice Centre is arguably one of Denton Corker Marshall's best in recent years, in terms of the economy of space planning, the creative composition of the volumes vertically with horizontal projections, the subtle and refined use of textures and transparency, all resulting in a building that is rather surprising and unprecedented. This building has received over 25 awards, including the RIBA (Royal Institute of British Architects) National Award for Architecture in 2008.

Dancing dragons (curves and figurative lines): Curving lines, surfaces and forms are used more prominently in Denton Corker Marshall's work after 2000. There is a variety of 'speeds' of these dynamic movements. There are slow and gentle curves, as in the waving outlines on the plan of a façade surface at the Victoria Gardens (Wave Apartments) and In Time Lotte Shopping Centre, both in Beijing, completed in 2003 and 2008 respectively. InTime Lotte Shopping Centre presents a glass wall waving in, out and around like a stage curtain, acting both as a reference to the old Peking Opera House which once stood on the site, and a visual communication point between inside and outside in the context of the everyday theatre of urban life.

Then there are fast and dynamic curves moving in three dimensions like dancing dragons, as found in the two competition entries, the Riverfront Cultural Park of Changsha (2004) and Jinsha Bridge of Hangzhou (2010). The Riverfront Cultural Park consists of a whole tube of rooms and spaces dancing and hovering along the riverfront, with a vertical cylinder tower behind. Jinsha Bridge is a three-strand twist of two curving bridges (for functional and recreational purposes) with the third dancing strand as a supporting structure (designed with Arup).

Another structure that fits in this category is the curve presented in a somewhat three-dimensional movement, but this time it is an outline of a flower petal. The Nanning Gateway sculpture (2000/2002) – a cluster of ten red metal sheets – petals – varying from 10 metres to 20 metres high, scattered on one side of a highway over a distance of 600 metres. They appear as a giant lotus flower gradually opening up to motorists driving past down to the city of Nanning, a flower that will 'blossom' again next time one drives past.

All these three cases are new to the 2000s. In the 1990s, the curves were largely horizontal and less dramatic, as in Melbourne Gateway (1995/1999) with its long orange wall waving gently along the highway; and South Bank Grand Arbor of Brisbane (1997/1999), where the curving path and tubular net insinuate horizontally. In the Wave Apartments and InTime Lotte Shopping Centre, the horizontal curve lines are extruded vertically as a curving surface of a large volume. In the two competitions for Changsha and Hangzhou, the curves accelerate to a new height in a daring three-dimensional articulation. In the case of the red flower petals of Nanning Gateway, a figurative mimesis is found, because of physical resemblance to a lotus – it is a new moment in Denton Corker Marshall's language at the edge of abstraction, as their work tended to be always abstract. Interestingly, the petals, each made of two metal sheets painted red, are abstract and figurative, structural and ornamental.[18] They stand at the edge of two realms, on the frontier of a new architectural language now on the rise – a

Riverfront Cultural Park Competition, Changsha, 2004.

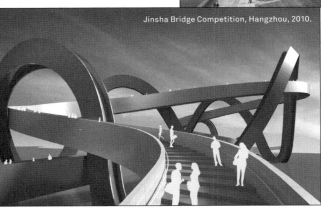
Jinsha Bridge Competition, Hangzhou, 2010.

InTime Lotte Shopping Center, Beijing, 2004/2008.

blurring between the two realms for a dynamic expression of emotion and life force. In this sense, all these cases, including the wave curtains and the dancing dragons, as well as the earlier experiments, are part of the same development towards a new awareness of life forms.

Sweeping horizontal lines: The use of horizontal lines as strictly horizontal or gently bent was evident in the Cape Schanck House of Melbourne (1997/1999) and the Federation Square competition (1997) respectively. As a concept, the 1986 Centrepiece was also forerunner of the use of extra-long horizontal lines. Yet the drama or the intensity of the sculpture is not materialised until in the 2000s, where we witness a clearer assertion of long horizontal lines as absolutely level, or as part of a sweeping slope of a vast rolling landscape. There is a Cartesian or Platonic simplicity, and a clearer sense of speed and eternity, evident in these long horizontal lines. They are designed to reflect, one may argue, not only the immediate landscape one sees, but also a new sense of the time, speed and geography of the 21st century, where we live 'here' or 'there', in a synchronized network, always staying connected instantly and constantly worldwide.

For the absolutely horizontal, we can enlist Medhurst House of Melbourne (2002/2008) and View Hill House and Vineyard (2010/2011). Both use a cantilever to project a platform or a tube, and both are self-consciously 'absolutist' in their horizontal stretch and in their orientation. View Hill House, for example, adopts exactly an east-west and north-south orientation, for the tube top and base respectively, in a Cartesian intersection. The cantilever projects the tube 6 and 9 metres at the front and back for the View Hill House, and as long as 11 metres for the Medhurst House – an astonishing feat achieved in the 2000s (using metal plates).

For the long and slightly bent 'slopes', the examples are Australian War Memorial (Canberra, 1999/2001), Stonehenge Visitor Centre and Museum (first scheme, Wiltshire, UK, 2001), and Zhongkun Vineyard and Hotel (design, Beijing, 2010). These works convey a sense of immensity of the great landscape, and a sense of speed and geographical stretch of the land and the sea we are flying across daily on IT networks.

...hurst House, Melbourne (Yarra Valley), 2002/2008.

Zhongkun Vineyard and Hotel, Beijing (Yanqing), 2010.

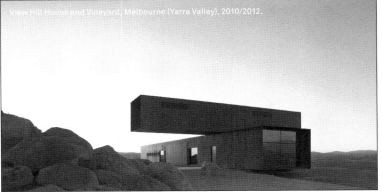
View Hill House and Vineyard, Melbourne (Yarra Valley), 2010/2012.

The Platonic cube: In the use of cubes, we can again find some earlier embryonic ideas evident in the 1990s and before, such as Emery Vincent Design Studio (Melbourne, 1986/1987) and Anna Schwartz Gallery (Melbourne, 2000/2000) where metal or concrete cubes are found as interior elements, or in Melbourne Museum (1994/2000) were larger cubes are found but in a more playful and dancing position. What emerges in the 2000s is a much stronger use of the cube or cubic form as a significant object in the whole composition, in scale dimension and in the posture adopted, as clearly Platonic or Cartesian – weighty, absolute and eternal, as if to register us once again to the ground in contrast to the new sense of time, speed and movement. In any case, the rise of these platonic cubes is clear in Brisbane Square (2001/2006), National Museum of China (competition, Beijing, 2004), Australian Embassy, Jakarta (2009/2015), Singapore University of Technology and Design (competition, 2010), and Australian Pavilion for the Venice Biennale (2011/2015). Judging from the frequency of the cube being employed in the decade, it surely plays an important role in the spectrum of the practice's work, as a new range emerging in the 2000s, a range that stretches from the Platonic to the lightness of speed and movement.

Australia in the world: As Denton Corker Marshall became more involved in overseas projects, there emerged a clearer understanding of home country in a global perspective. The practice's work in the 2000s exhibits a new clarity on what may constitute an identity of Australia and how to represent it in architecture. The aspects that constitute this national identity are immensity and mineral richness, which are represented in a scale-less form and in the metallic textures of the form respectively. And the form is often the Platonic cube or a cubic box. While the austere cube or box, with minimal detail, refers to an immense openness of the land and the sky of Australia, the metallic textures on the box refer to the mineral wealth. Of course, an obvious example is the Australian Pavilion for the Venice Biennale (2011/2015), a bold and monumental cube, austere and scale-less, suggestive of the immense landscape from which it comes, literally and symbolically. For the surface texture, the box is clad with South Australian black granite, creating a pavilion that is mysterious, pristine and grand, with layers of

National Museum of China Competition, Beijing, 2004.

Singapore University of Technology and Design Competition, 2010.

meaning to be uncovered for a visitor. Another example is the Australian Embassy in Jakarta (2009/2015), where five different metallic textures are applied to five cubic blocks – zinc, copper, brass, steel and aluminium – to suggest the wealth and the scale of the land they represent.

A typology for the millions (China): There are two ways of dealing with scale or magnitude in Denton Corker Marshall's work of the 2000s: a scale-less object suggesting immensity and openness, and a form of literal quantity with thousands of entities or units of space. The latter is in fact a typology for tens of thousands of homes with often a million square metres. This typology or form of real quantities is found in the office's work in China. With the framework approach ('sketch') which leads to a powerful urban design outcome and delivers the huge quantity that allows local variations and contributions from the local DIs, the design framework is open, tolerant yet still highly controlled as a work of the Melbourne office. One of the smart strategies is well designed repetition. The result is enormous developments in more than ten cities across China, with some thirty five thousand homes, or several millions of square metres now completed. Denton Corker Marshall has provided two major contributions. The first is order and identity given to these massive quantities, so that there is a formal discipline: Suprematist strips and planes, distinctive colours as identities for parts or blocks, and an overall formal control on massing along with spatial planning with repetition. When these measures are applied consistently, they assist these developments to achieve a clear formal character and identity. The Sunshine 100 housing developments across these Chinese cities have acquired iconic imagery.

The second contribution in these developments, which is more important as it captures the entire spatial form, is an urban typology constructed for these thousands of homes as a collective entity in or around the cities. The typological approach of the 1970s-80s, never abandoned as a spatial idea, has been used to consciously form defined residential and urban communities, with boundaries, entries, axes, visual corridors, shifting levels, blocks as 'walls', with an essential idea of a court or plaza (as found in European civic spaces and Chinese compounds large or small), and spatial order created from repetition.[19] This is one of the strongest legacies of post-modernism of the 1970s-1980s as explored by

Australian Pavilion, Venice Biennale, 2011/15

(Photomontage: a, in Australia; b, in Venice).

Rossi and Krier, and much tested in Denton Corker Marshall's work. It made a brief appearance in the Beijing Embassy (1982/1992) and then reappeared at a much larger scale for a more direct contribution to China in the 2000s. This typological approach is useful because it offers a design method to counter the destruction of urban textures by the onslaught of modernist forces of singular towers in a universal open space. And this is particularly useful for China today, as the destructive tendency is so strong and sweeping.

As to the examples, the best cases should include Southbank International New Town of Chongqing (2003/2007), and Euro-City Plaza of Nanning (2001/2007). The Shangdong International New Town of Nanning (2005/2011, with 1220 homes at 186,000m^2) is a good example as well, of the formation of enclosed urban communities with entries, internal laneways, shifting levels, community gardens, and a meandering long block as a 'wall' defining the realm at the back.

Denton Corker Marshall's work in China is not limited to housing developments, yet these huge projects constitute the majority of the practice's work there. Compared to the practice's work in many other places, especially in Australia and the UK, these huge buildings in China are relatively broad and at times crude, a result of the quantity, the limited time, the budget, and the second-tier market targeted. Yet the urban typology, the formal order and the visual identity of these developments has been clearly achieved, with the method of the 'sketch' or the framework approach that is tolerant yet controlled. It is a complex process of 'transfer' of design knowledge, but, in the process of localisation, it is also a form of interaction and synthesis, of a formal order from the west with the dynamic forces of the market of Asia.[20]

Accelerated Modernism and the Case of China

For the practice's work as a whole, because of an accelerated pace of work and culture and a faster means of communication worldwide, combined with closer global interactions with Asia and Europe from Australia, the Suprematism of the 1990s has developed into a globalist perspective in Denton

Euro-City Plaza, Nanning, 2001/2007.

Shangdong International New Town, Nanning, 2005/2011.

Corker Marshall's work in the 2000s. It has three important characteristics: a tolerance of global diversity of design speeds, from the slow and refined work of the UK to the fast and relatively rough work of China; a new awareness of Australia and its immense landscape, with the use of mineral texture, horizontality and the scale-less cube; and a broader spectrum of work from the dynamic (3-D movements, vertical assemblage, a fast form for the millions) to the eternally Platonic, a spectrum that is new to the decade, even though it has evolved from the past. An important background for the generation of these new ideas and tendencies is the speed and pressure of the new time, for which China appears to have been a contributing factor, for the immediate output of the practice in the past ten years.

To assess the relevance of China in Denton Corker Marshall's overall profile, we may look at it in two aspects: generic relevance and direct involvement. For the generic relevance, we may argue that there is an implicit influence or impact from the speeding-up of the work in China on the office's overall outlook, contributing to an overall global backdrop of acceleration. As the frontier of development in global comparison, China is arguably where speed, acceleration and the pressure to act and innovate quickly are felt most acutely, although of course the tendency is a global phenomenon. Secondly, for direct involvement, out of the immediate pressure and demand from this market, Denton Corker Marshall has made at least four contributions: the residential typology for the millions; a dynamic formal experiment with 'dancing dragons'; a return to culture and a rise of smaller, more refined work concerning tradition and landscape; and a flattening of barriers between China and the West by addressing similar design ideas simultaneously in China, the UK and Australia.

The 'typology for the millions' and 'dancing dragon' have been explained above. The third contribution is a return to cultural issues, after the design of the Beijing Embassy (1982/1992), and an emergence of smaller but more refined work addressing cultural tradition and the experience of a vast rural landscape. The cases here are the InTime Lotte Shopping Centre glass curtain scheme in central Beijing (2004/2008) referring to the memory of an old Peking Opera House on site, and the Zhongkun Vineyard and Hotel (design, 2010) which responds sensitively to the grand and rolling landscape to the northwest of Beijing at the foot of great mountains, for its creation of a poetic experience in the fields with the long horizontal walls.

The fourth contribution is a breaking down of barriers between China and the West, putting the work in China onto a global platform of design discourse, by addressing similar ideas in China, the UK and Australia simultaneously. For example, the idea of vertical assemblage of boards and boxes were explored in Beijing and Manchester, for the Zhongkun Tower and the Civil Justice Centre, respectively, at about the same time (2002/2007). Similarly, the use of long and horizontal lines to respond to the landscape were explored in the Vineyard and Hotel project in Beijing (2010), View Hill House and Vineyard in Melbourne (2010/2012) and the Stonehenge Visitor Centre Scheme One and Two in the UK (2001 and 2009/2014). In these cases the work in China enters a global platform of design discourse across different countries. Of course this is a general trend, and the works in China, by architects Chinese or not (such as Wang Shu, Yung Ho Chang, Cui Kai, Urbanius, Zaha Hadid and Rem Koolhaas), are increasingly entering into the global discourse. Denton Corker Marshall's cross-border transfer of ideas is part of this development. Specifically, Denton Corker Marshall's overall contribution includes four aspects with the first one, a fast form for the millions; the typology of New Towns and City Plazas; in the provinces across China, as arguably the most distinctive.

Comparing the office's work in China, the UK and Australia, and judging from the highly refined work accomplished in Melbourne and Manchester especially, it must be acknowledged that the work in China so far – though vast in quantity – is less refined in aesthetic quality in terms of formal and tectonic design. Given the quality of the work produced by Denton Corker Marshall elsewhere, there is a vacuum to be filled – more work should be expected of Denton Corker Marshall in China with a more refined quality and a more articulated theoretical position, as China itself is increasingly interested in 'innovation' in new development. To innovate one surely needs to encourage more out of design, discourse and creativity.

With a population 20 per cent of the world's total, an economy the second-largest, and a development rate among the world's highest, China is emerging as a special 'plateau' with its own 'intensities' (to borrow words from Gilles Deleuze and Félix Guattari), in developing a new modernity for the millions, with its scale and pressure, and perhaps its own ethics and aesthetics – a modernity that is hitherto unknown. China's contribution in design discourse is yet to emerge, but is likely to arise and take shape in the next few decades. How to comprehend and engage with this China, and how to forge and help develop a new discourse with China's participation, is arguably one of the most important issues we are facing today in the discipline of architecture. In this sense, Denton Corker Marshall's work in China, as a transfer of design knowledge, but also an interaction and the forging of a synthesis, between a formal order from Australia (and the Anglo-western world) and the dynamic forces of the market and society of China, is one of the most important adventures undertaken in our time.

Jianfei Zhu is Associate Professor in architecture at the University of Melbourne, and is a supervisor of PhD and other research students.

The author wishes to thank John Denton and Greg Gong from Denton Corker Marshall in Melbourne for their generosity in sharing information and reflections, and the staff of Sunshine 100 in China, Yin Xiaojun, Zhou Xiang and Huang Jian of Nanning, Wang Rong of Chongqing, Tang Xinyu of Beijing, and Zhang Jie and Bai Jinhu of Tianjin, for their generous support in my field trip in June-July 2011. This paper is part of a project 'Australia-China Collaboration in Urban and Architectural Design', funded by Research Collaboration Grants of the University of Melbourne (2010-12).

1. Haig Beck and Jackie Cooper (eds) Australian Architects: Denton Corker Marshall – A Critical Analysis, Canberra: Royal Australian Institute of Architects, 1987; Haig Beck/Jackie Cooper, Peter G. Rowe and Deyan Sudjic, Rule Playing and the Ratbag Element: Denton Corker Marshall, Basel: Birkhäuser, 2000; and Leon van Schaik (ed.) Non-Fictional: Denton Corker Marshall, Basel: Birkhäuser, 2008. There is also a Chinese publication in 2002: Jianxin Nie and Xiangqing Chen (eds) Aodaliya DCM Zuopin Shilu (Works of an Australian architect DCM), Beijing: China Architecture and Building Press, 2002. See also Denton Corker Marshall, expresso < expressway: Denton Corker Marshall's non-architecture, Melbourne: Denton Corker Marshall, 2006.
2. These two years, 1985/88, represent the year designed and the year completed, here and thereafter in the article, unless otherwise indicated.
3. Beck and Cooker, Australian, pp. 23-6, 31-6, 37-42 and 51-4, for the 222 Exhibition St, 101 Collins St, State Library and Princes Plaza respectively. For the two office towers, see also Beck/Cooker et al, Rule Playing, p. 116.
4. Beck/Cooper et al, Rule Playing, pp. 86-7, 90-5. See also Jianfei Zhu, 'Denton Corker Marshall in China: Interactions', in Leon van Schaik, Non-fictional, pp. 134-40 and Jianfei Zhu, 'Export or Dialogue', Architecture Australia, vol. 99, no. 5, Sept-Oct 2010, 97-8.
5. John Denton, 'Interview', Melbourne, 9 Nov 2012; see also Beck/Cooper, Rule Playing, pp. 88-89, 96-107.
6. Haig Beck and Jackie Cooper, 'Development of a Formal Language', in Beck/Cooper, Rule Playing, pp. 33-7.
7. Beck and Cooper, 'Development', in Beck/Cooper, Rule Playing, p. 33.
8. Beck and Cooper, 'Development', in Beck/Cooper, Rule Playing, p. 34.
9. Beck/Cooper, Rule Playing, pp. 40-63, 188-9, 194-209, 223 and 232-37.
10. Leon van Schaik, 'A Tale of Twined Cities', in Schaik, Non-Fictional, pp. 8-23.
11. Information provided by the office of Denton Corker Marshall in 2005.
12. Based on my interviews with the Sunshine 100 staff in Nanning, Chongqing, Beijing and Tianjin, June-July, 2011.
13. Information provided by the DCM office in 2007 and 2011.
14. According to the DCM office in 2007 and 2011.
15. John Denton, 'Interview', Melbourne, 9 Nov 2011.
16. Beck and Cooper, 'Development', in Beck/Cooper, Rule Playing, pp. 34, 34-7; and Peter G. Rowe, 'Rule Playing and the Ratbag Element', in Beck/Cooper, Rule Playing, pp. 19, 13-32.
17. Beck and Cooper, 'Development', in Beck/Cooper, Rule Playing, p. 34.
18. For a discussion on the 'mimesis' of the City Gate project, see Zhu, 'Denton', in Schaik, Non-fictional, pp. 134-40 and Zhu, 'Export', Architecture Australia, pp. 97-8.
19. For a discussion of the embassy, see Zhu, 'Denton', in Schaik, Non-fictional, pp. 134-40 and Zhu, 'Export', Architecture Australia, pp. 97-8.
20. The idea of 'transfer' of skill and knowledge in the process is mentioned a few times by John Denton (John Denton, 'Interview', Melbourne, 9 Nov 2011).

访 谈

约翰·丹顿，
安尊·菲茨杰拉德
访谈录——
聂建鑫先生

聂建鑫： 中国建筑工业出版社在 2002 年，即中国建筑设计行业十年繁荣时期的起点，推出了一本专门介绍丹顿·廊克·马修建筑事务所作品的专著。我们就从这段时间丹顿·廊克·马修建筑设计事务所完成的一些优秀作品谈起吧。

安尊·菲茨杰拉德： 十年前您写的那本介绍我们公司的专著收录了 25 个项目，其中 3 个在中国，当时这 3 个中国项目都仅仅是方案。十年前，在中国，除了澳大利亚驻北京大使馆外我们没有任何建成的项目。现在出版的这本书将介绍我们的 31 个项目，其中 12 个是中国项目且 6 个已经建成，还介绍了 3 个已建成的住宅项目，3 个城市规划以及基础设施建设项目，其中南宁的城标已建成。如今，尽管那些参与竞标的文化公共建筑作品没有建成，但是一些办公楼和商业设计项目已建成投入使用，比如北京的乐天银泰百货和中坤大厦等项目。我想这体现了最近十年间的变化。

约翰·丹顿： 这些年来，我们在中国各地有了一系列的住宅开发项目，体现出一些特别的设计理念。这些住宅开发项目的设计过程中，我们确立了一整套关于如何在中国进行设计工作的主张，以此为出发点，我们做了不少项目。但是，最引以为傲的还是我们的公共类建筑，尽管到目前为止我们只有一些方案可展示。举例而言，位于北京西北的落樱酒庄，它包括了高品质的葡萄园、酒庄以及一座精品酒店；或者回顾我们之前的竞标项目，例如长沙滨江文化中心，竞赛时我们是第一名，不过该项目最终委托给了国内一家设计院。但是机会不断，我们现在就正着手设计一些新的项目，我们期盼这些建筑能够建成。

安尊·菲茨杰拉德： 我们知道建筑师多通过设计公共建筑而得到业内认可。但即便如此，在中国，商业和办公类建筑的市场非常大，不管是斯蒂文·霍尔（Steven Holl）还是大都会建筑事务所（OMA）或任何建筑师都在此领域中竞争。我们观察到，中国的建筑设计标准在三个方面迅速提高：一是设计构思，二是建设过程，三是建成效果。我认为从精工细作的角度看，像落樱酒庄这类项目颇具潜力。大量客户在促进着建筑设计这个新兴市场的发展，其中涉及的中产阶级就可能达到两亿。这一新兴的群体开始追求质量，而过去人们则更多追求数量。我们在这一大环境下，力求满足数量的需求——我们设计并建造了两三万套公寓。而跨入这个新的时代对我们来说是中国给我们的一个振奋人心的挑战。

聂建鑫：您提到建立一套关于如何在中国进行设计工作的主张，能深入谈谈这些主张以及您所目睹的过去十年间中国建筑和建筑设计领域发生的变化吗？社会和自然环境有什么变化？

约翰·丹顿：正如安尊前面提到的，过去十年中中国建筑市场发生了重大的变化。20世纪80年代我们在中国建造了我们的第一个建筑作品，从那时开始到90年代发生了很大的变化，而90年代到21世纪间的变化则是天翻地覆。建筑的数量以及类型迅速增加，建筑的质量不断提高——不论是大批量的建筑市场，还是公共建筑项目这样的高端建筑，均是如此。

安尊·菲茨杰拉德：我们看到的中国的不凡之处是其工业革命的速度。在西方国家，工业革命历时一个世纪；中国只用了几十年就完成了工业革命进入到科技时代，而建筑的发展也与时俱进。中国过去的工业产出以世界标准衡量微不足道，而如今，其工业产量仅次于美国，位居世界第二。建筑设计方面也经历同样的飞跃性发展，最近荣获普里兹克建筑奖（Pritzker Prize）的中国建筑师王澍就是一个很好的例子。可见中国建筑的发展是多么迅速。约翰，回想早期的经历，与设计院的合作……

约翰·丹顿：与早先的设计院相比，中国本土建筑师的情况如今大有不同。不管这是否因为海外教育的经历或其他什么原因，现在中国本土有着优越的建筑设计环境。我们总是说我们会一直向中国提供技术转让，直至中国自己的技能水准与国际接轨，但这个过程不会很长。我认为我们的估计是正确的，过去估计这个过程会是十年，但实际上可能要经过十五年，我们现在正处于这个过程之中。

安尊·菲茨杰拉德：可喜的是，中国正在出现一批具有世界建筑设计水准的高质量建筑，例如内蒙古鄂尔多斯博物馆或是广州歌剧院。中国的一些建筑的施工水平可以与澳大利亚和欧洲媲美。

约翰·丹顿：同时，中国一些新兴的建材企业使用的是世界最先进水准的设备，比如当今世界上最好、最先进的玻璃制造厂就在中国。南方玻璃厂生产的玻璃比匹兹堡或任何地方都更好，拥有更好、更新的设备。南方玻璃厂隔壁的玻璃热弯厂是全世界最好的热弯厂之一，它的设备先进程度极高——最初是为了服务国际市场，如今也能同时满足国内需求。在中国，技术方面的提高非常快，这意味着建造能力不再受到限制。在中国，我们有着和世界上任何地方一样的自由和契机。而且，总体规划和设计的水准也越来越高，过去形式僵化的网格式布局已被多样化所取代。

安尊·菲茨杰拉德： 我们承担各种规模的项目，无论是量身定做的小型别墅还是城市中更大区块的设计。我们非常乐于接受城市环境所带来的挑战，可以说是致力于城市发展的专业人士。毋庸置疑，中国是个正在城市化的国家，这本身就创造了诸多挑战，而这些挑战为人们带来了各种环境问题。我认为问题在于如何在环境和发展之间寻求平衡⋯⋯

约翰·丹顿： 是的，我认为中国仍然存在自己的问题，这从北京污染问题和各地的各种问题中都可以看到。很多重大问题仍待解决。但是中国的现代化进程速度飞快，我认为中国人的环境意识日益增长。对于可持续发展的问题和所有这类问题，中国学习得很快，并且会通过政府机制推进，所以我们有理由期待快速的变化，期待人们认识到可持续发展的重要性。中国关注着世界其他国家的发展，并同时做出反应——动作非常迅速。

聂建鑫：您提到与中国建筑设计事务所和设计院合作的经验。在和中国的设计伙伴合作的过程中，您运用了一些什么技巧可以使双方交流顺畅，以保证取得最好的工作成果？

约翰·丹顿： 回想设计澳大利亚驻北京大使馆的经历，那时确实有很多问题。当时普遍缺乏技术——没有施工技术，没有工程计划系统技术，本地的建筑行业几乎不具备任何技术，所以建造使馆遇到很大的困难。而现在这种情况已经完全改变了，这也是我们刚才谈到的巨变之一。现在中国建筑行业有相当的技术水准，足以建造复杂的幕墙系统结构，或其他需要新技术，包括新的电脑系统等技术来完成的复杂建筑。建筑行业的迅速变化中正包括了这些方面。

建大使馆的时候，我们有用之不尽的人力，但是他们都来自农村，都没有经过技术培训，来北京后就住在施工现场，一年两次回家收割庄稼。尽管中国现在仍然有技术短缺的问题，国家仍然不断培训工人，使他们能够完成更具技术含量的复杂工作，但我认为巨大的变化已经发生，实实在在地发生着⋯⋯应该说，我们也懂得在建筑设计上因地制宜，这方面我们经验丰富。对我们来说，要运用智慧建成开发商想要的建筑，不会把开发商当成傻瓜，当然他们也不会接受傻瓜的做法——不要过度夸张，而是在处理问题中运用智慧。

安尊·菲茨杰拉德： 关于技术进步的问题，以赫尔佐格和德梅隆建筑事务所（Herzog & de Meuron）设计的鸟巢为例，现场焊接技术是多么精良，还有最后各配件的组装，已经达到了极高的精度。至于和当地设计院的关系，我想我们向他们介绍了一些技能使他们在技术上得以提高，我认为这是我们为技术输出所做的努力。

约翰·丹顿： 我们在中国二、三线城市做了很多项目，那里的技术水平相对要低些——尤其是设计院的技术水平不是很高。因此，我认为，合作可以使他们在建筑设计的内容和方式方面得到技术上的提高。我们所做的工作中一大部分是根据二、三线城市的施工水平实现我们的建筑设计方案。这不仅是努力引导他们的过程，也是在现实环境条件下力图达到最佳建设成果的过程。而像上海、广州、北京和深圳这样的大城市则有着较高的技术水准，这些地方拥有所需的技术，也拥有技术水平最高的工人。

安尊·菲茨杰拉德： 有趣的是，在金沙湖步行桥设计竞赛方案中，我们设计的桥梁的连续拱结构（和 Arup 合作），创造了一座具有强烈动感的雕塑般的作品。遗憾的是这个项目没有实施，设计竞赛最终被取消了。但是最近我去昆明的时候（2013 年初），发现昆明机场采用了与我们当时的设计非常类似的结构，不同之处在于它是采用混凝土结构完成的，而且已建成。让我非常诧异的是，在这个地震带，结构工程师们能够提出这种设计方案并且能够实施，这要是在过去是不可能的，因为结构工程师极其保守。

约翰·丹顿： 回到十年前建南宁城标（Nanning Gateway）的时候，南宁没有能力做。当时在上海切割、预制并按国际标准制造城标的结构组件是完全可能的，最终他们在上海加工好组件后，用军用火车运到南宁。当时中国也具备一定的技术水准，但是仅限于一些大城市，因此很多问题都和地点有关系——大城市中技术发展最快，人们需要最先进的办公大楼和功能最完善的歌剧院，等等，然后这些技术才逐渐地被引进到二、三线城市。这是一个发展的过程，但是现在这个过程进行速度非常快，这点很关键。

安尊·菲茨杰拉德： 现在的建筑工业国际化程度很高。比如，我们设计的悉尼科技大学的计算机和工程信息系馆，建筑立面是一个复杂的铝板幕墙系统。它的板材来自德国，运到上海切割、加工，在悉尼进行氧化处理，然后运往布里斯班制作，最后运回悉尼安装。在整个生产过程中涉及了三个国家和五个不同的城市。

我们一直在做大量的海外工程，这种跨文化的工作经验要求我们的思维方式多样化。尽管我们的设计注重广泛的环境背景，但是我们会根据项目所在地的情况和环境提供适宜的方案，因为项目可能在各种不同的国家和城市，可能是在堪培拉，可能是在曼彻斯特，也可能是在上海或雅加达。

约翰·丹顿：我明白你所说的，但我不完全同意你的说法。比如，我认为人们对曼彻斯特民事法院感兴趣是因为其设计理念和处理方法是非英国本土的，它来自澳大利亚。我们的设计理念在一个国际化的平台上展现，同时在这个国际舞台上产生积极的影响，在中国也是这样。我们与其他海外建筑师一起，对中国建筑业的发展产生着影响。我们是影响中国发展的海外力量之一。我认为我们的作品反映了我们独特的风格。

安尊·菲茨杰拉德：但是如果与那些仅在本土做项目的澳大利亚建筑师交谈，就会发现他们的思维有一定的模式……我清楚地看到我们思考问题的方式不大一样，因为我们是在全球范围工作，承接世界各地的设计。

聂建鑫：当效率、经济性与设计冲突时，你怎样平衡它们之间的关系？

安尊·菲茨杰拉德：这是每一个建筑师都不可避免要遇到的问题——各种法规的要求，开发商需要赚取利润，对建筑成本的控制……建筑师的问题是在众多的制约条件下如何创造出优秀的设计作品。每当你看到一件杰出的建筑作品，你总会很佩服建筑师们可以克服各种条件的影响。当然，拥有一个好的客户有助于好的设计作品的实现。

约翰·丹顿：是的，我想这在不管哪个国家或是任何地方都是一样的。能否设计出好的作品取决于客户的水平，你和客户之间的关系，以及你们双方是否拥有共同的目标并能够实现这一目标。如果你拥有一个好的中国客户，你就有了可能做出好的作品的前提条件，但这仍旧是和人相关。最终，所有的建筑设计都还要取决于客户和他们是否乐意为建筑作品支付相应的设计费用。我想中国也不例外。今天在中国有一个现象，很多人追求有品位的作品，在这么大的一个国家里这样的人不会少。问题在于如何使这个人群成为你的客户，以及你如何满足他们的需求。

聂建鑫：全球金融危机后的国际形势是否影响到你们对追求卓越的决心？相比之前你们完成的像墨尔本博物馆、墨尔本会展中心和墨尔本城标这样杰出的作品，金融危机是否影响到甚至减少了你们在城市建设复兴方面创作优秀又有影响力的设计作品的机会？

安尊·菲茨杰拉德：据我观察，从建筑设计的角度来看，全球经济危机前建筑行业处于过度做作的状态，明明是一些小项目,却在追求并设计极其奇怪的形状和外观——那段时间就如同是巴洛克时代的建筑过度矫饰时期。而现在经济形势给建筑业带来了一些约束,我认为这对建筑设计行业是一件好事。

约翰·丹顿: 你将它归为一个原因,或是它就是导致这种情况出现的本原? 因为在我看来建筑业正在经历一个阶段,我也期待今天人们认识到那个阶段是一个"过头"的时代。现在的建筑设计行业所受的约束变多了,这可能和导致全球金融危机的因素有关。

安尊·菲茨杰拉德: 合理性一直是我们在设计中注重的问题,是功能布局的基础。我们的建筑设计从基本入手,显而易见的正门入口、各个功能空间的组织以及道路导向等等,都以合理性为基准,然后再结合鲜明的外观特点。我们不会为了标新立异而设计奇特的形体外观,再把功能装进去——对我们来讲,这种做法从根本上是和优秀建筑设计的宗旨背道而驰的。

约翰·丹顿: 现在仍然有很多人还在做这些"过头"的设计。像迪拜这样的地方,它们经历过严峻的经济衰退期,因此过度设计的现象有所减缓,但是我认为仍然有不少的建筑师在过度设计的死胡同里义无反顾地向前。

聂建鑫: 最近几年来,丹顿·廓克·马修建筑设计事务所的国际项目数量在公司整个业务量中占到较高的比例,体现了贵公司的全球化进程。从市场角度来看,是否有些国家较之其他国家具有较强的建筑设计市场? 都有哪些国家?

约翰·丹顿: 如果从施工质量角度来看,世界各地的标准大相径庭。中国是个日益增长的市场,施工质量标准一直在提高。但如果你要看建筑施工在哪里完成,那么在澳大利亚,你可以建一幢不错的建筑;在欧洲,特别是在北欧,你可以建一幢非常好的建筑;在日本,你可以建一幢非常精美的建筑。这完全和这个国家历史上的建筑方式、建筑标准和技能相关。有的地区在建筑方面经验丰富,很显然,在这种环境中从事建筑业更好。但是情况总是在变。在意大利,过去可以找到非常好的石匠,但是现在就比较困难,因为很多优秀的意大利石匠都去了北欧,现在从事这项工作的是一些技术不大精湛的移民工人。技术受经济利益的驱使,欧盟各国人员流动较大,因此技术也外流。日本一直保有很高的技术水平,但是日本是个非常稳定的社会,不可能有大量日本工人到处流动——他们不会到处跑,社会较为封闭。当日本发生建筑行业过热的状况时,开始出现很多非日籍劳工(大多是非法劳工)参与施工的情况,施工标准有所下降,然而日本经济发展速度再次放缓,且历时多年,所以施工标准降低的压力不复存在。

安尊·菲茨杰拉德: 这个问题的另一方面是不同国家对设计的重视程度,比如北欧的斯堪的那维亚一贯对设计质量高度重视。在亚洲,最近几年中新加坡在这方面开始崭露头脚,例如我们的新加坡亚洲广场(Asia Square)项目,其中包括了双子塔和城市客厅空间。这个项目在以下两点的共同作用下得以实现:一是采用了市区重建主管机构(URA)的规划导则,二是客户购买相邻的两个地块给我们提供了创造新型城市空间的机会。我们和市区重建主管机构协商争取在地块上除了按照要求设计的人行横道外,设计一个名副其实的巨大的城市空间,面积约6000m^2、高达18米的空间完全向公众开放。这是一个非常独特的城市建设成果。建造类似这样的城市空间曾是很多人的梦想,但未曾有人能够将这一梦想变为现实。现在,这个梦想在新加坡实现了。我想同样的机会在中国也会出现,因为在那些政策开明的城市政府和开发商协同努力,他们希望在自己的城市里有世界一流水平的建筑。

聂建鑫: 您如何看待全球化的现象,以及它对本土建筑表现形式的影响?例如,澳大利亚驻北京大使馆反映了中国传统住宅的建筑风格,而你们其他的中国项目在风格上倾向于国际化,您如何看待这个问题?

约翰·丹顿: 设计的国际化受媒体、媒体的变化以及人们观察事物方式的共同作用。随着中国的开放,中国人看到了世界上其他国家的发展情况,人们接触到一些有意思的想法,自然会去追求。我认为国际化通常对本土建筑传统是有害的,但是本土建筑传统也需要有适应能力。今天不会有太多的人想去修建一个中国传统胡同了,就像我们也不会想要再去修建一套乔治亚王朝时代(Georgian)或其他什么类似风格的建筑了。事物的本质即是如此,而现代媒体更是加速了这一变化。我相信,中国优秀的年轻建筑师正以审慎的态度对待中国的传统建筑,并将对中国传统建筑思想的理解融汇进现代建筑的设计当中,而且已经有一些非常精美的建筑出现。

安尊·菲茨杰拉德: 是的,作为澳大利亚人,我们没有很深的文化渊源,因此我们也就没有什么伟大的传统。我们都是移民,最多也就六、七代人,因此我们已经习惯于接受和传播新的理念。这就是我们工作的方式。尽管弗兰姆普敦(Kenneth Frampton)和一些人提倡批判性的地域主义是正确的建筑形式,我认为这个理论不一定正确。在澳大利亚,有人会说,这座昆士兰(Queensland)北部风格的建筑非常符合当地特点,但是我不会苟同这种看法。我的立场是澳大利亚所有的建筑都是"进口"的——不管是乔治亚王朝风格的，还是新古典主义或者国际式，都是在其他国家发展深化的理论中衍生出来的建筑风格，然后成为澳大利亚设计理念和建筑理念的组成部分。近些年来我们在文化上成为一个越来越多元化的国家。今天，来自120种不同文化的人们共同构成了多元文化的澳大利亚，这是我们文化的特殊性。

我想正是因为这个缘故，我们在其他文化环境中工作比较容易。而来自单一文化——例如意大利、日本或其他国家的建筑师可能会觉得较为困难。我们的另一个优势是我们都说英文，而英文又是国际商务通用的语言。

约翰·丹顿： 回想过去，我们在北京设计大使馆项目的时候，正是中国大量拆除传统民居建筑的时候——胡同被大量拆除，取而代之的是一些质量不高的预制混凝土高层建筑。我们将大使馆视作是关于胡同背后传统理念价值的实验。大使馆不是胡同的复制品，它的设计吸取的要素包括了传统的围墙，南北向的轴线、建筑坐北朝南的布局，以及适应当地气候环境特点和建筑特点的方窗和空间。北京是一个"墙"的城市，如城墙、紫禁城的围墙和内城的围墙乃至于长城。但是，我们的建筑完全采用现代手法设计，因此它不是"中式建筑"，但是具有中国式的理念，这一点是澳大利亚驻北京大使馆所特有的。在我们的其他中国项目中，这样的理念较少呈现，因为没有必要或合适的机会来采用这种方式和语汇，更多的项目都是大型住宅开发，客户需要的是新的、国际化的设计作品。我们仍然尝试运用色彩等设计元素，我们使用红色，它代表幸运；我们也使用黄色，我们认为这是把皇家专用的颜色民主化，过去是禁止民间使用这种颜色的。

安尊·菲茨杰拉德： 我们非常喜欢运用色彩。我们自己有一套相对清晰的设计语言，并自如地运用在设计中。有人说这是一种风格，我个人不知道这是不是一种风格——我们不一定这样定义它，但是也许你可以说它就是一种风格。

约翰·丹顿： 我们其实采用比较纯粹的，更为晚期现代主义的观念来看待事物，因此这种观念具有很强的建筑学属性，因为这着眼于材料的属性、各元素功能之间的搭配以及建筑建构的逻辑。我们基于这几点来设计建筑，我们不会专注于创造建筑风格。

安尊·菲茨杰拉德： 我们一直对现代艺术、抽象派较有兴趣，所以它们一直存在于我们的意识之中。我们的很多建筑外形都很简单，由一组独立的元素组成，要么是一个简洁的细长立柱外形，要么是一片薄如蝉翼的构件。整个建筑看起来如同是由一些分散的元素构建而成的。我们不会把建筑外形和材料堆集在一起,而是通过苦心经营找出如何利用凹进和分割处理手法,以展现纯净形体和元素。

约翰·丹顿: 是的,我们展现这种建构的原因是,我们乐于表达建筑是如何配搭的以及如何实现清晰的构造和清晰的理念。

聂建鑫: 让我们谈谈丹顿·廓克·马修作为一个事务所的发展历程吧。三位创始人现在是否仍然担任主创设计师?新董事以哪些主要方式对公司发展贡献才能?丹顿·廓克·马修设计事务所的设计风格是否得到传承?到什么程度?新董事是如何将这种风格具体体现出来的?

安尊·菲茨杰拉德: 公司的现况是大多数高层人员都已经在这里工作了很长很长时间,比如我自己是1982年加入的,已经参与设计超过30年了。因此我们有一种延续性,我们自己不太看得出来,但是很多来自其他公司或其他环境背景的人来到我们的工作室,都会提到一种非常明显的"丹顿·廓克·马修"设计思考方式的存在。我觉得这很好,我们的工作室存在一种独特的设计文化气息。

约翰·丹顿: 当然三位创始人中的两位仍然工作在第一线,但是其他各位董事,如安尊所说,是公司长期发展的中坚力量。

安尊·菲茨杰拉德: 我们其中一位新董事是龚耕,他是中国人。在我们公司,中国项目的重要性众所周知。龚耕是公司的合伙人,主持和负责公司对华业务。龚耕以及公司会说普通话的员工可以和客户直接沟通,这对公司业务帮助很大。有趣的是,这对我们公司的运营还有一个好处:中国的投资者们日趋成熟,并力图在海外市场分散投资风险,他们来澳大利亚购买开发房产,这又给我们在澳大利亚带来了一个新市场。

约翰·丹顿: 长期的海外实践,尤其是东南亚地区的工作经验,使我们对当地文化颇有了解,比如语言。因此,在中国项目的设计工作中,据我所知,我们与那些我所知道的其他任何公司都不同。从我们公司出去的所有设计文件都是中文的,而不是英文的。我们公司有这个语言实力,所以我们和中国的客户可以保持非常好的关系,我们公司有说中文的员工——我们想方设法使他们的语言优势最大限度地发挥。

安尊·菲茨杰拉德：人们经常问我们这样一个问题，即我们什么时候去中国设立分公司。我们的回答是不在中国设分公司。原因就和丹顿刚谈到的内容有关，即我们控制设计以及保证长期维持的设计质量的能力。这些年来我们发现最好的方式还是由墨尔本（总部）进行设计工作。我们过去有过不少分部，但是我们逐渐关闭了这些分部，原因有两个：一是一个大型机构运作的长期管理问题，这成为影响设计工作的一大工作内容；另外一个原因是，把核心设计力量集中起来对于保持质量很关键。当然，我们会因为没有当地办事机构而失去一些中国项目的设计机会，但是我们绝对相信，这种工作模式会使我们能够为中国和所有其他国家的客户提供我们最好的设计作品。

聂建鑫：在过去十年内，贵公司设计的哪一座建筑充分体现了丹顿·廓克·马修建筑设计事务所的设计特点？为什么？

约翰·丹顿：曼彻斯特民事法院——这是一个海外的项目，设计工作在墨尔本完成，是通过设计竞赛获得的项目。我们最好的设计项目大概都是通过设计竞赛赢得的。

安尊·菲茨杰拉德：这个项目得到了广泛的认可，获得了20多个国际奖项……

约翰·丹顿：不少是建筑设计类奖项和环保可持续发展奖项。这个建筑的质量和施工非常好，体现出了所有我们在建筑中希望体现的元素，它被誉为建筑界一项重要的作品。

安尊·菲茨杰拉德：我同意。曼彻斯特民事法院是目前为止我们建筑设计水平的最佳体现，从平面到剖面都清晰反映了内在的逻辑；还有在建造过程，建筑形式的表达，以及一个外立面的环保外罩如何作为城市里的一件艺术品得以展示等等方面，都是如此。我认为在许多层面上，这座建筑都达到了很好的效果。一个建筑师在他一生的创作生涯中可以达到如此高度的作品可能只有寥寥几个。这是目前我们最好的代表作。我们希望下一个最优秀的作品会在中国产生！

**INTERVIEW WITH
JOHN DENTON
AND
ADRIAN FITZGERALD
BY JIAN-XIN NIE**

Jian-Xin Nie (JN): China Building Industry Press commissioned a book about Denton Corker Marshal in 2002, on the cusp of China's 10-year architecture boom. Let's start by talking about some of the significant Denton Corker Marshall projects built during this period.

Adrian FitzGerald (AF): The book of our work, published 10 years ago, featured 25 projects, three of which were in China, all unbuilt. In this monograph there are 12 China projects out of a total of 31, six of those built. I think that demonstrates the shift in the past 10 years, from virtually nothing built in China other than the Australian Embassy. While we haven't built the cultural and civic projects – they've been competition entries – we do have the office and retail projects such as InTime Lotte and Zhongkun Office Tower. The book includes three built residential projects, and of the three urban design/infrastructure projects, the Nanning Gateway was completed.

John Denton (JD): We have a really positive stream of residential development that we've done across China, which has particular philosophical ideas about architecture. In this residential work we have built up a whole set of propositions about how we work in China, and in that process turned out a lot of work. But the works that we tend to be most proud of are the institutional ones and for those we really only have unbuilt examples to date. Some of them, like the Zhongkun Vineyard project, which is a high quality vineyard, winery and hotel project northwest of Beijing; or looking back to our unbuilt competition work, such as the Changsha Riverfront Cultural Centre, a project that was a competition which we believe we technically won but then went to a local firm on reference to the mayor. But there are other potential opportunities, such as the things that we're working on now. We look forward to the opportunity to build those.

AF: We know in architecture you get your recognition out of the civic buildings. However having said that, in China there is a really strong emerging architectural scene in the commercial sector, whether it's Steve Holl, OMA, or whoever. The difference we have observed is that the architectural standards in China have just gone up exponentially, in three ways: one, what is being conceived; two, being built; and three, being built well. I think in terms of finessed works, the Zhongkun Vineyard project has potential. There are clients that are pitching to a market that has emerged, the middle classes of which there is maybe 200 million. This is a new affluent group that wants quality, when in the past it's all been about quantity. We've been part of that treadmill, churning out the quantity side – the 20 or 30 thousand apartments we've actually designed and built – that's the exciting challenge for us about China, moving into this era.

JN: You mentioned you developed a whole series of propositions around working in China – can you tell us a bit more about that, and a bit more about the changes you have witnessed in the Chinese building and architecture industries over the past 10 years? What about social and natural environments?

JD: Well, certainly, as Adrian said the last 10 years have seen major change in the construction market. We first built in China in the 1980s, and the change from the 1980s to the 1990s was radical, and the change from the 1990s to the 2000s was radical. It continues to change exponentially in terms of output and types of buildings being built, and the construction quality that you're seeing, either in the general construction market or in the specific higher-quality end of the market, such as civic projects.

AF: What we generally observe that's extraordinary about China is the speed of its industrial revolution. In the West it took a century; in China it took a couple of decades. The country has gone through an industrial revolution, into a technological stage, and the architecture is moving along with it. To begin with, China's industrial output in world terms was tiny, and now, it is second only to the US. In architectural terms, it has gone through a similar set of leap frogs, evidenced by the fact we now have a Pritzker Prize winner in China – Wang Shu. This tells you how rapidly it's changing ... because those early days, John, with the design institutes …

JD: There's a massive change from those early days with the design institutes and with local Chinese architects, whether it be because of overseas training or whatever, so there is now a very strong local architectural environment. We'd always said that we would be doing technology transfer until such time as Chinese expertise took over, and it wasn't going to be long – I think we were right. We might have thought it would happen in 10 years; the reality may be it will happen over 15 years, but essentially, it's happening.

AF: The exciting thing about that, in world architectural terms, is that there are quality buildings – look at Ma Yansong's Ordos Museum in Inner Mongolia, the distinctive shaped one, and the glass reinforced concrete shapes now being built, or the carved stone in the opera house at Guangzhou. They're now constructing buildings in China that are comparable to Australia or Europe.

JD: At the same time in China they're building new industrial enterprises that include some of the world's best quality equipment. So now the most up-to-date, best glass production factory in the world is in China. South China Glass is better than Pittsburgh or wherever – it has newer, better equipment. The glass-bending factory that is next door to China Glass is one of the best in the world, so there's sophistication in the factories, which were initially built to service the international export market, but are now servicing home as well. The skills are growing very fast. It means you're not hamstrung in what you can build. You've got the same sorts of freedom and opportunity that you have elsewhere in the world. Also the sophistication of master planning has become more significant … because over the years the master planning has been very grid-like and basic in its formality, and it's now allowing itself to become a little bit more sophisticated.

AF: We've always embraced scale, at all levels, from the very small bespoke house to shaping large sections of cities. We've certainly always enjoyed the urban condition as a challenge. We're committed urbanists, as it were. China is certainly urbanised, and that brings a lot of challenges in itself, which leads into the question put to us about the environment. I'd like to think that the question being posed is about the balance of environmental conditions versus developmental conditions …

JD: Yes, I think China still has issues, and you can see it in the pollution problems in Beijing, you can see it in all sorts of ways right through China. There are still some massive problems that the country has to get through. But China is modernising at a great speed, and I think there's a growing awareness of the environment in China. China's very quick to pick up on things like sustainability, and all those sorts of issues, and start to implement them through its own government structures, so you can expect reasonably fast change and you can expect sustainability to become significant. The country is observing what the rest of the world is doing and reacting accordingly at the same time – it's a very fast-moving sort of thing.

JN: You mentioned working with the Chinese architectural/design institutes. What are some of the techniques you've developed to assist smooth communications and ensure the best possible outcomes when working with your design partners?

JD: When we look back at the Australian embassy in Beijing, it's true to say there were some problems at the time. There were enormous skills shortages – no construction skills, no programming skills, virtually no skills whatsoever available in the local industry, so it was very difficult to build. That's just changed completely now, again part of the radical change we've been discussing. You do now have a fair level of skill in the construction industry, which is now able to build sophisticated things like curtain walls, or complex things that require new techniques and new computer-based systems and things like that to achieve them. And that's just reflected in the speed of change.

With the embassy, there was an endless supply of workers, but they were unskilled, and twice a year they went home for the harvest, because they were just straight off the farms. They'd come to Beijing, live on site, and then they'd go home for the harvest. I think China still has a skilling problem and the country is still trying to train the workers to be able to do the more skilled things, but that is happening, it's definitely happening ... I should say, as well, that we're also experienced at building within the means. To us, it's building intelligently to get the outcome that the developer wants, and not killing him with silliness – not that he would have accepted it anyway ... not trying to overdo it, but trying to be clever about the way we did it.

AF: In terms of developing skills, if you take, as an example, Herzog & de Meuron's Birds Nest Stadium, the on-site welding is extraordinarily good. And the final coming together of the pieces, the tolerances they had to work within ... In terms of our relationship with the design institutes, I think perhaps where a skills transfer has occurred is in small ways where we have introduced things that may have helped them up-skill.

JD: We've done a lot of work in second- and third-tier Chinese cities, and there the skills aren't so good – particularly in the design institutes – so we have worked with them, and I think we've participated with them in up-skilling what they do, and how they do it. So by and large most of the work we've done is about building within the capacity of that second-tier city's builders. It's a process of trying to guide them, but also to work within the limits of what's available. The real skills always tend to reside in the large cities like Shanghai, Guangzhou, Beijing and Shenzhen, and that's where the capacity lies, and where you can get the most skilled workers.

AF: Interestingly in our scheme for the Jinsha Bridge Competition, we designed the loops (with Arup) to hold up the bridge, creating a very dynamic sculptural object. Unfortunately, it didn't go ahead, the competition was abandoned. But I was just in Kunming recently (early 2013) and was intrigued to see that the new Kunming Airport uses a very similar concept of looping structure, albeit in concrete, and it's built. It surprised me that in an earthquake zone, the structural engineers are prepared to sign-off on such a design, which hasn't been the case in the past; they've been extremely conservative.

JD: Going back 10 years to the Nanning Gateway, it couldn't be built in Nanning. But they could cut and prepare and manufacture the gateway elements to international standards in Shanghai, so it was done there and brought down on an army train. There were skills then, but only in the big cities. So a lot of it is to do with location – the take-up of skills happens fastest in the big cities, where the people want the most sophisticated office buildings and the most sophisticated opera houses, etc. Then the skills flow on slowly into the second- and third-tier cities. So it's a process, but the process is happening very fast, that's the key thing.

AF: And we're in the process of the globalisation of the construction industry. For instance in our Broadway Building at the University of Technology, Sydney, where we have a complex aluminium screen cladding the building, the aluminium is sourced in Germany, it's then sent to Shanghai for cutting and fabricating, then it's shipped to Sydney for anodising, and then it's trucked to Brisbane for fabrication, and then it goes back to Sydney for installation. So there's three countries and five different cities all involved in its production.

The reality is, we do a lot of work overseas, and working cross-culturally makes you think in different ways. While we've always been contextually based, we've always responded quite directly to a project's situation and circumstance, but it could be in a variety of locations – it could be in Canberra, it could be in Manchester, it could be in Shanghai, it could be in Jakarta, or one of the many other cities we've worked in.

JD: I hear what you say, though I don't necessarily agree with you. I think, for example, that people find Manchester Civil Justice Centre interesting because it has a thought process and an approach that's not from England – it's from us here in Australia, we're putting the ideas out internationally. We're influencing what happens, we're having an impact on what happens internationally, and that's what happens with China as well. We're part of the overseas influence that's impacting on the way China's developing. I think our work is uniquely our work.

AF: But in conversations with other Australian architects, who only work in Australia, they think in certain ways ... it's clear to me that we think in other ways, and that's because we are working globally, in all kinds of locations.

JN: How do you balance efficiency and economy when they conflict with design?

AF: Well that's the perennial question for any architect ... how to achieve a great outcome with the inevitable restraints of regulations, the requirement for the developer to make money, to design within budget, and so on. And whenever you see a great work of architecture you're always impressed that someone was able to cut through all of that. Clearly, having a good client is a great benefit to realise great designs.

JD: Yes, and I think that doesn't change wherever you are, regardless of the country. It's the quality of the client and the quality of your relationship with the client and the objectives that they, and you, have and are able to implement. If you've got good Chinese clients, that allows you the potential to do good work, but it's all still related to people. All architecture is dependent on your client and your patronage, and I think China is no different. But China exhibits the fact that there are a lot of people keen to do unique and interesting things, and it's a very big country; therefore there are a lot of people like that. It's a matter of how you tap into those people and satisfy their desires.

JN: What about the global situation post-GFC – did it affect your determination for excellence? Did it impact on or reduce opportunities to create excellent and influential work, particularly in terms of

urban revitalisation, as you have done with projects like the Melbourne Museum, Exhibition Centre, and Gateway?

AF: What I observed architecturally prior to the GFC was an excess in architecture, as in people designing the strangest shapes and expressions for fairly insignificant projects … it was almost a baroque era of excess. And the fact that there's now a degree of restraint because of the financial situation … I think that's a plus for architecture!

JD: So you attribute that as the reason, or is it simply contributing to the reason? Because it does seem to me that architecture is just going through a phase, and I'd like to think those excesses are now being seen for what they are: excesses. And there is more restraint now. Perhaps the GFC has had something to with that.

AF: Our work always has rationality. It has always been grounded in sensible, functional arrangements. All our buildings work in fundamental ways, from readily finding where the front entrance is, to moving through it, to wayfinding, etc. And then we've had a strong expression coming with the functionality. But we've never just done strange shapes for the point of doing strange shapes, and then somehow squeezing a function into it, which to us seems fundamentally at odds with what good architecture is about.

JD: There are still a lot of people doing very excessive things, but some places … like Dubai, have gone through a major downturn, and therefore the excesses of building in those sorts of locations has slowed, but I still think there are plenty of architects trying to pursue the dead end of excess …

JN: In recent years, Denton Corker Marshall's overseas projects account for a higher percentage of its total work, which demonstrates the process of the firm's globalisation. From a market point of view, do some countries have a stronger architectural/design market than others? Which ones?

JD: If you look in terms of quality of construction, then different parts of the world have completely different sorts of standards, and China is a growing market, where the standards are improving all the time. But if you look at where you can get a building built ... in Australia you can build a fairly good building ... in Europe, in particular Northern Europe, you can build a pretty good building, in Japan you can build an exquisite building. It's all to do with the history of how the country builds and the standards and skills they have. Obviously, it's nicer to build in environments where you have experience of building very well. But it fluctuates. In Italy, you used to be able to get great stone work, but now it's harder, because a lot of the good Italian stonemasons have gone to Northern Europe, and now there is less-skilled immigrant labour doing the work. The skills go where the money is, and within the European Union there's a lot more movement of people, so the skills have gone elsewhere. Japan has always retained very high skills, but then Japan's a very static social fabric where there is no big likelihood that masses of Japanese workers are going to go somewhere else. They don't. It's closed. What's more, when Japan had an overheating in the building industry and the country started to get a lot of non-Japanese labour coming in (mainly illegally) on building sites, the standards dropped a little bit. But Japan's slowed down again, for many years now, and that hasn't been a pressure of late.

AF: Another side to this question is commitment by various countries to design. Scandinavia, for example, has always had a strong commitment to design. Within Asia, a place like Singapore has emerged in recent years. I am thinking in particular of our Asia Square project in Singapore, the twin towers with the big City Room space. This was a coming together of certain guidelines from the Urban Redevelopment Authority (URA), and an opportunity created when the client bought the two adjacent sites. We negotiated with the URA to create more than just the required pedestrian laneway, to instead create a really substantial civic gesture for the city, a 6000m^2, and 18m high space dedicated to the public realm. It is quite a unique urban achievement. Many people have dreamed of creating spaces like this but not been able to build them, yet now it's realised in Singapore. And I think there are going to be similar opportunities in China, in enlightened cities with mayors or councils or developers coming together who want to lead and aspire to achieve the world's best.

JN: What are your thoughts on globalisation and its influence on local building language? For example, the Australian Embassy in Beijing reflects a traditional Chinese residential building style; your other Chinese projects tend to be more international in style.

   JD: The internationalisation of design is a function of media, and the changes in media, and the way things are seen. And as China opens up and sees what's happening elsewhere in the world, the people see ideas that they find interesting and seek them out. I think that internationalisation is usually to the detriment of local building tradition, but then local building tradition has to be adaptive anyway. You don't want to build a traditional Chinese Hutong anymore. You don't want to build a Georgian something-or-other. That's just the nature of things, but it's exacerbated by and increased in speed through new media. I'm sure the better young architects in China are thoughtfully considering Chinese traditions and building contemporary buildings with an understanding of the Chinese traditions. And there's some very fine work that's done there.

   AF: Yes, but as Australians we don't have a long culture. So we don't have any great traditions. We are all migrants in one form or another, for six or seven generations at longest. So we're used to importing and exporting ideas. And that's a way of working. Whereas Frampton and others would propose critical regionalism as being the appropriate form of architecture, I don't think that's necessarily correct. There are some in Australia who would say there's a north-Queensland style of architecture that's appropriate for that region … but I don't think we subscribe to that view. My proposition is that all the architecture in Australia is imported – whether it is Georgian, or neo-classicism, or internationalism … they are all building styles derived from theories developed elsewhere, and then exponents of the theory designed and built them within Australia. Culturally, in recent years we have become very diverse. Whether it's the fact that we've got 120 different cultures that make up current-day multi-cultural Australia, that's something special that we have. I suspect because of this we find it easy to work in other cultures, whereas architects coming from more singular cultures – Italian or Japanese or whatever – might find it harder. We also have an advantage in that we speak English and that's the international language of business.

JD: Going back for a moment, the Australian Embassy in China was built at a time when China was demolishing its traditional residential fabric – the Hutongs were being cleared out and the country was building rather bad precast high-rise housing. And we saw the Embassy as being a didactic conversation about the values of the traditional ideas behind the Hutongs. It wasn't a copy of it. It was about using a containing wall; having a north-south access, buildings facing south from the north end of the axis; using poched out squares and spaces [wall niches] – and that's all to do with weather conditions and environment, and a function of the environment and a function of the way Beijing was, with its many walls (the city wall, the wall around the Forbidden City, the walls around the inner part of the city, etc). But our building was done in a totally contemporary way, so that it was not Chinese, but it reflected Chinese ideas that were inherent in our Beijing Embassy. The other Chinese projects that we've done tend to have less of that; there's not the same sort of need or opportunity to be didactic in that sort of way because it's been about building big residential developments and what the client's want is what's new, what's international … we've still tried to play with things like colour, using red, which is a lucky colour, and we also used yellow, which we saw as democratising the imperial colour, the use of which was forbidden by the Emperors.

AF: We enjoy colour. We've always had a relatively clear design language that we've developed and are comfortable working with. People describe it as a style, I don't know … we don't necessarily see it in those terms but I suppose you could.

JD: We have quite a pure, sort of late-modernist view on things which is therefore very architectural because it's about materiality, it's about the way things go together, it's about the pragmatics of architecture and how to make it work. We make architecture out of that; we don't make it as a sort of 'style construct'.

AF: We've had an interest in modern art, abstract expressionism, so it's always there in the back of our heads somewhere. Many of our buildings are elemental, in the sense of being composed of an assemblage of elements, be it a stick, or a blade. The architecture reads as being made up of discrete elements. We don't smash forms and materials together; we go to a lot of trouble to articulate how pure forms and elements are expressed via re-entrants and separation.

JD: Yes we articulate the 'how' because we like the idea of expressing how the building goes together; the clarity of the tectonics and the clarity of the idea.

JN: Let's talk about Denton Corker Marshall's progression as a firm. Are the three founding members still chief architects in design? What are the main ways the newer directors contribute to the firm's output? To what extent is the original Denton Corker Marshall style being carried forward, and how do the newer directors embody this style?

AF: The reality of the make-up of the firm is that most of the senior people have been in the practice for a long, long time. For example, I joined in 1982, and I've been part of the design process for over 30 years, so there is a continuum. We don't see it so much, but certainly people coming to our studio from other offices and other environments often comment on the presence of a consciously 'Denton Corker Marshall' way of thinking about design. And I think that's right. There is a distinctive culture of design in the studio.

JD: And certainly two of the three founding directors are still going strong, and the other directors, as Adrian has said, are part of the long-term progress of the firm.

AF: One of our newest directors is Chinese, Greg Gong. This is a clear acknowledgement within the firm about the importance of China: Greg is a full director, actively running the practice's China projects. Greg and our Mandarin-speaking staff's ability to communicate directly with the clients is a big plus. Interestingly, that's working for us in a secondary way, with Chinese investors buying and developing in Australia, as they're maturing and wanting to spread their investment risk in other markets. That's an interesting new angle for us.

JD: Because we've worked overseas for a long time, especially in South-East Asia, we've understood about things like language. So working in China I think we're different from just about any other firm I know in that all of our design work leaves here in Chinese and not in English. We actually have the skills here, and therefore we have a very easy relationship with clients in China and there are people here who speak Chinese – we actually make a significant effort in maximising how it works.

AF: One of the questions we're perennially asked is when we are going to set-up an office in China. And our answer is: we're not. The reason being, in relation to what John's talked about, is our ability to control the design and maintain the quality of design that we've always produced. We've found that over the years it's best to do it out of the Melbourne/head office. We've had regional offices in the past and we've progressively closed them due to two reasons: one, because of the ongoing management issues in running a big organisation, which becomes a sideline of the main game, and two, the key thing is to have our core design team all in one location. So of course we're going to miss certain projects in China because we don't have a local office, but we absolutely believe that we will produce our best work for China – and for all countries – by staying with this model.

JN: Which building produced in the past 10 years best represents Denton Corker Marshall design characteristics, and why?

JD: Manchester Civil Justice Centre: because it's an overseas building, designed here, won in competition. And our best work has probably come via competition wins.

AF: It's been acknowledged, it's won over 25 awards …

JD: Architecture awards, sustainability awards. It was very well built, it reflects all the things that we aspire to in architecture, and it's acknowledged as being a significant piece of architecture.

AF: I agree. And it's a culmination of a lot of ways of thinking about design and architecture for us, in terms of the clarity of the plan, which also reads through into the section, and how it's built, the expressiveness of its forms, and how the environmental veil on one façade presents as a piece of art to the city … on so many levels I think it works extremely well. There are only a handful of these projects that you might ever achieve in your working career. It's currently our best representation. But we hope our next best will be in China!

CUL-
TURAL

文化 + 公共建筑

+CIVIC

曼彻斯特
民事法院
MANCHESTER
CIVIL
JUSTICE
CENTRE

通过多轮国际设计竞赛，DCM 最终赢得了英国曼彻斯特民事法院的项目。该法院是英国西北部的司法总部，15 层高，建筑面积达 3.4 万 m^2，设有 47 个法庭、75 个法律咨询室以及办公区域和附属空间。曼彻斯特民事法院位于斯宾菲尔茨区，这个地区由于极富创新精神和可持续性的设计，成为曼彻斯特城区的大型旧城复兴区。| 在平面上，法庭和办公室按线型排列，由法庭的大小和数量决定每一层线型结构的长度；各层中的线型结构分别向建筑两端伸展并错位变化，形成动感的造型，使建筑呈现出虚实相间的效果。这些元素共同在侧立面形成了个性鲜明的建筑外观；而在两端的立面上，则营造出光影互动的效果，同时具有深度和内涵，显示出雕塑般的力量。| 营造透明体的建筑手法隐喻着法庭不再是戒备森严的地方，而是一个平等和向公众开放的场所。| 曼彻斯特民事法院为寻求正义的人们提供帮助，它被英格兰和威尔士最高法院首席法官誉为"一座代表着民事司法正义新纪元的伟大建筑……可能是现代最好的民事和家事法庭"。

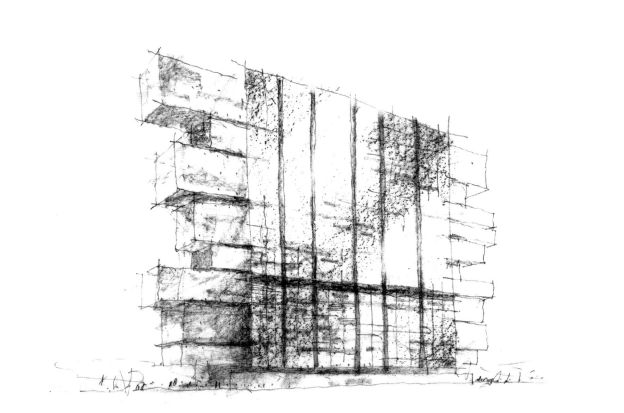

An international competition-winning design for the headquarters of the Ministry of Justice in north-west England, the complex provides accommodation of 34,000m² over 15 levels. It houses 47 courtrooms, 75 consultation rooms, offices and support space. It is located in Spinningfields, the large-scale regeneration area in central Manchester marked by innovative and sustainable design. | The working courts and offices are designed as long rectilinear forms, articulated at each floor level, and projecting at each end of the building as a varied composition of solid and void. The hierarchy and number of courts set the length of the 'sticks'. In side elevation, these elements collectively establish a dynamic and distinctive building profile; in end elevation, they form a powerful sculptural interplay of light and shade, depth and complexity. | The architectural implication of the transparency of the building is that the courts are not forbidding or concealed, but open and accessible. | Welcoming and comprehensible for people seeking justice, the Manchester Civil Justice Centre is described by the Lord Chief Justice of England and Wales as 'a magnificent building representing the start of a new era for civil and family justice … probably the finest civil and family courthouse of the modern age'.

建设地点	英国 曼彻斯特
客　户	皇家法院服务署
	伦敦地产联盟
荣获奖项	曼彻斯特法院中心项目荣获超过 25 项奖项，其中包括：
	– 2007 年澳大利亚联邦建筑师协会约恩·乌松最杰出国际建筑奖
	– 2008 年英国皇家建筑师协会国家建筑奖
	– 2008 年英国皇家建筑师协会英国可持续发展设计伙伴奖

Location   Manchester, United Kingdom
Client   Her Majesty's Court Service
            Allied London Properties Management Ltd
Awards   The Manchester Civil Justice Centre has won over 25 awards including:
            - Jørn Utzon Award for the Most Outstanding Work of
              International Architecture, Australian Institute of Architects, 2007
            - National Award for Architecture, Royal Institute of British Architects, 2008
            - English Partnerships Sustainability Award, Royal Institute of British Architects, 2008

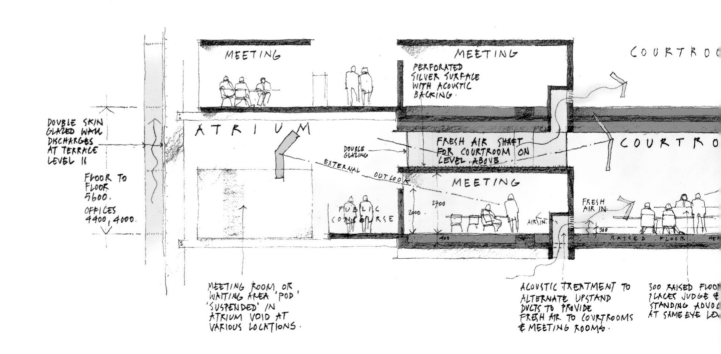

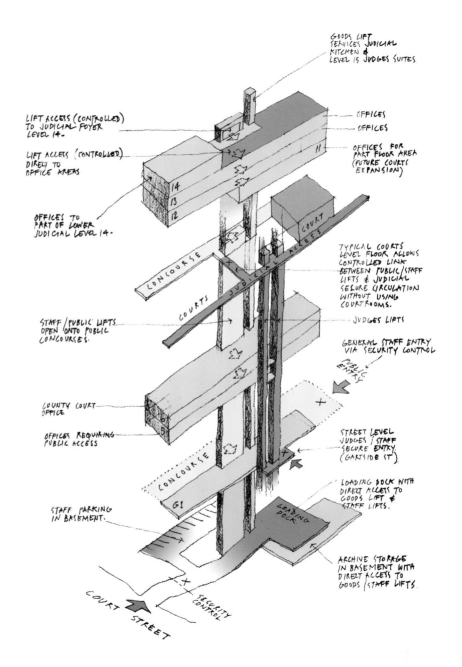

英国
巨石阵
游客接待中心
STONEHENGE
VISITOR
CENTRE

英国巨石阵是全世界最为重要的远古文明遗址之一，每年有将近一百万游客前来参观。修建巨石阵游客接待中心的计划旨在为这座珍贵的古代遗迹恢复其往昔的尊严。| 这项计划包括若干项内容，包括完善游客服务设施，以及为巨石阵和周边景点提供更好的解说场所，最重要的是改善更大范围内的自然环境，为游客更好地欣赏巨石阵提供一个最佳环境。| 接待中心建筑在构图上简洁而独特，既与周围环境协调，也强调了遗址的重要性。游客服务设施采用银色金属屋顶，由细长倾斜的立柱支撑。屋顶的边缘采用穿孔设计，使得下方的展厅和教育设施内部光影斑驳。展厅和教育设施分别设在两个独立的盒子中，其中一个盒子采用玻璃，另一个采用木材。| 游客接待中心位于巨石阵西部2.4km，隐藏在巨石阵看不到的位置，从游客中心到巨石阵有专门的车辆接送游客。

coloured alternative to overlay on base landscape — Canopy B, Boale C.

Stonehenge is one of the world's most significant archaeological sites, and receives close to one million visitors each year. The scheme for a new visitor centre restores a sense of dignity to the treasured ancient monument. | Fulfilling several important aims in the management of the World Heritage listed site, the scheme includes the construction of improved visitor facilities, better opportunities for interpretation of the Stones and the wider site and, most importantly, a substantially improved landscape setting in which to appreciate Stonehenge. | The architectural composition of the centre is simple yet distinctive, sensitive to its surroundings and to the significance of the monument. Supported by slender angled stick columns, a sliver metal roof shelters various visitor amenities. Its edges are perforated to dapple the light reaching the exhibition and education facilities housed below. These are housed in a pair of single-storey cubes – one glass and the other timber. | Sited 2.4km to the west of the Stones, the visitor centre will not be visible from the monument. A transit system will transport visitors to the Stones.

coloured alternative to overlay on base landscape — Canopy B, Boall C.

modulating roof canopy above the landscape.

glass block + timber block – placed beneath the canopy.

建设地点	英国 威尔特郡
客　户	英国历史文化遗产保护机构
荣获奖项	2009 年巨石阵游客接待中心国际竞赛概念邀请赛头奖
	巨石阵游客接待中心国际竞赛概念邀请赛头奖
	（方案于 2007 年被新方案取代）

Location Wiltshire, United Kingdom
Client English Heritage
Awards - First prize, Stonehenge Visitor Centre,
limited international ideas competition, 2009
- First prize, Stonehenge Visitor Centre & Interpretive Museum,
limited international ideas competition, 2001 [scheme abandoned in 2007]

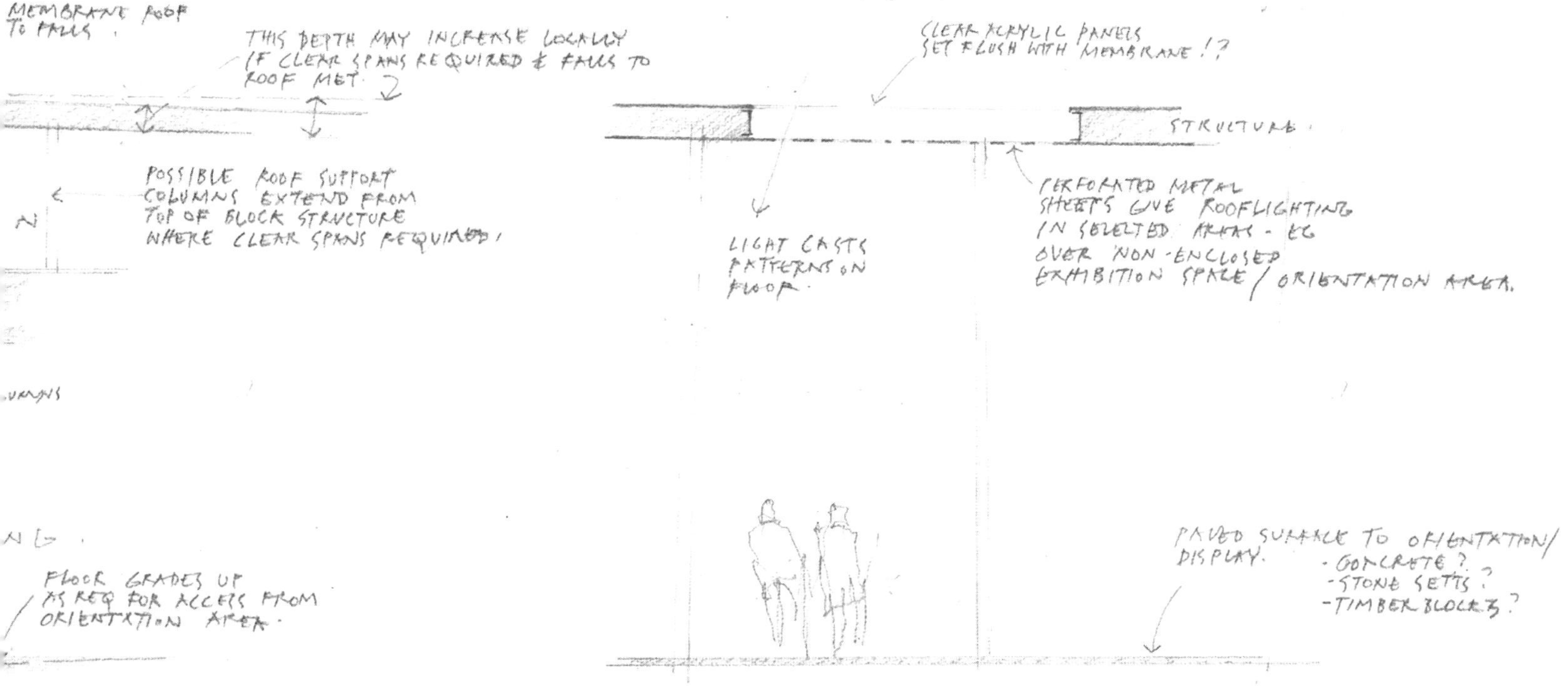

MEMBRANE ROOF
TO FALLS.

THIS DEPTH MAY INCREASE LOCALLY
IF CLEAR SPANS REQUIRED & FALLS TO
ROOF MET.

CLEAR ACRYLIC PANELS
SET FLUSH WITH MEMBRANE!?

STRUCTURE.

POSSIBLE ROOF SUPPORT
COLUMNS EXTEND FROM
TOP OF BLOCK STRUCTURE
WHERE CLEAR SPANS REQUIRED.

LIGHT CASTS
PATTERNS ON
FLOOR.

PERFORATED METAL
SHEETS GIVE ROOFLIGHTING
IN SELECTED AREAS - EG
OVER NON-ENCLOSED
EXHIBITION SPACE / ORIENTATION AREA.

FLOOR GRADES UP
AS REQ FOR ACCESS FROM
ORIENTATION AREA.

PAVED SURFACE TO ORIENTATION/
DISPLAY.
- CONCRETE?
- STONE SETTS?
- TIMBER BLOCKS?

1:50. — APRIL 09. — SECTION THROUGH ORIENTATION / EXHIBITION.

悉尼科技大学
计算机和工程信息系馆
FACULTY OF ENGINEERING
AND INFORMATION TECHNOLOGY
UNIVERSITY OF TECHNOLOGY
SYDNEY

悉尼科技大学计算机和工程信息系馆坐落在繁华的市区，是悉尼科技大学的新校区乃至悉尼中心商业区南端的地标式建筑。| 系馆 12 层高，附 4 层地下空间，其中包括技术先进的讲堂、办公室、研讨室、教学和研究实验室以及学生会、餐饮休闲区和自行车机动车停车场等功能空间。| 建筑外观如同一座造型非凡的雕塑，这一设计使之在周围传统外观的建筑中脱颖而出。建筑周身由四片倾斜扭曲的金属网板包裹。金属网板采用铝合金材质，其网眼组成的图案是"悉尼科技大学计算机和工程信息系馆"的二进制代码。| 狭长的中庭连接教学和社交场所，有力承载着综合教学的学院文化。建筑内部有敞开式楼梯、不规则的天桥和自由的社交场所，交通流线沿楼层的边缘而设。室内采用不规则素砼板和耐候钢等材料，体现出原生态的美学追求，营造出一种类似仓库的效果，使得学院更具有创意产业的特征。而除了连接首层各个空间外，这一采光性很好的中庭还连通了大学和周围的社区。| 计算机和工程信息系馆也称为百老汇大楼，它是悉尼科技大学打造一个标志性建筑区的梦想的一部分。这一梦想将耗时十年，斥资十亿澳元，并将成为悉尼城市中心区更新的一个重要部分。

On a prominent urban site, the Faculty of Engineering and Information Technology building creates a gateway to the university's revitalised city campus and the southern end of Sydney's CBD. | Twelve levels, plus four below ground, accommodate state-of-the-art lecture rooms, academic offices, seminar rooms, teaching and research laboratories, student union, and food and recreation areas, with bicycle and car parking. | The building is expressed as a singular sculptural object, setting it apart from the more traditional architectural expression of its neighbours. Four tilted and skewed plates envelop the building's volume. The plates are made of aluminium sheets perforated in a pattern derived from the binary code for 'University of Technology Sydney Faculty of Engineering and Information Technology'. | The faculty's collaborative learning culture materialises in the ultra thin crevasse-like atrium which links all teaching, learning and social spaces. It's a dynamic space occupied by open stairs, random bridge links and lounges for informal encounters, with circulation along its edges. The raw aesthetic achieved internally with materials such as off-form concrete and Corten steel engenders a warehouse quality, aligning the faculty with environments favoured by the creative industries. | Offering naturally-lit access through the building at ground level, the crevasse directly links the university to the local neighbourhood. | Also known as the Broadway Building, the Faculty of Engineering and Information Technology is part of the university's vision to deliver an iconic and connected campus, and a 10-year, $1 billion redevelopment that will help transform the southern approach to the Sydney CBD.

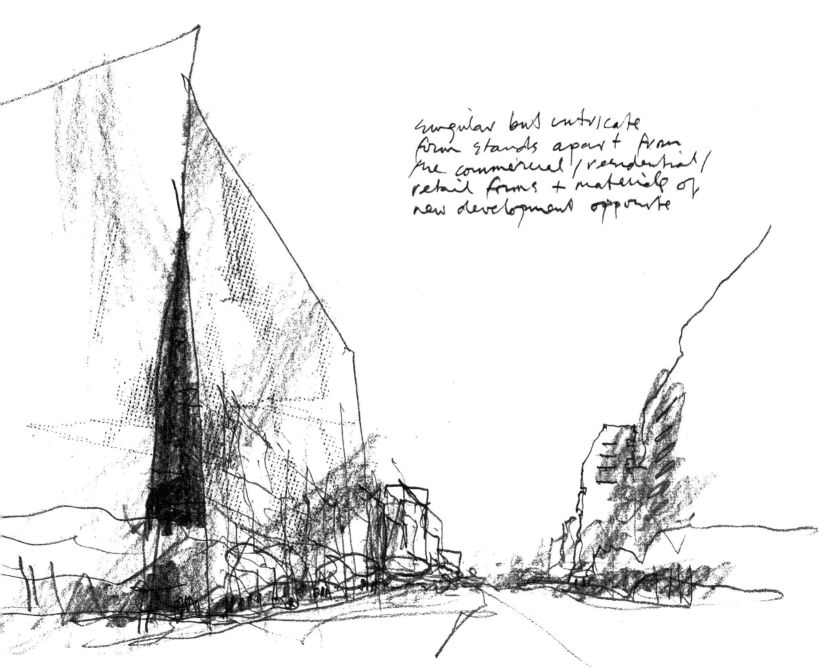

建设地点	澳大利亚 悉尼
客　　户	悉尼 科技大学
荣获奖项	2009 年悉尼科技大学百老汇大楼国际竞赛决赛头奖

Location	Sydney, Australia
Client	University of Technology, Sydney
Awards	First prize, two-stage Broadway Building UTS international design competition, 2009

新加坡
科技设计大学
SINGAPORE
UNIVERSITY
OF TECHNOLOGY
AND DESIGN

新加坡科技设计大学的设计是一次国际邀请赛的竞赛作品。该方案中，建筑面积为 21.3 万 m^2，采用创新的设计，在各建筑之间运用连通的结构，有力地体现了团队合作精神。在建筑的内部设计出可以灵活布局的空间，从而将普通的教室变成多功能建筑，可以根据不同教学需要灵活变换。自习空间则散布在建筑内的主要交通节点周围。| 校园中心完全由一个立方体组成，同时具有标志性和功能性。立方体内部设置了图书馆、礼堂、国际设计中心和一个光照充足、设有天窗的绿色空间；外皮采用预处理的钢件遮阳板。它作为标志性建筑，坐落在立柱雕塑群中间。| 为适应热带气候特点和满足可持续发展的要求，设计采用了一系列的重要手段，例如在整个校园范围内统一使用的智能遮阳外皮（包括重要的第五立面，即屋顶）以及最大限度地运用自然通风等。

Bold interlocking forms that powerfully express the spirit of collaboration feature in this limited international design competition entry for a 213,000m² university campus. Within the forms, flexible spaces turn conventional classrooms into hybrid buildings, which can be modified for a variety of pedagogical uses. Informal learning lounges are scattered around key circulation nodes. | The symbolic and functional hub, the Campus Centre, is composed of pure cubes. It comprises the library, auditorium, International Design Centre and a generous light-filled, green space with a skylight roof. Wrapped in pre-weathered steel sun-screening, the cubes sit as signature buildings on a sculptural field of pillars. | Essential design responses to the tropical climate and sustainability imperatives include a smart shade skin wrapping the entire campus (including the critical fifth elevation, the roof) and maximum use of naturally ventilated space throughout.

建设地点　　新加坡
客　　户　　新加坡 科技设计大学
合作建筑师　CPG 咨询公司

Location Singapore
Client Singapore University of Technology and Design
Associated Architects CPG Consultants

174　CULTURAL + CIVIC

印度尼西亚大学
中央图书馆
UNIVERSITY OF
INDONESIA
CENTRAL LIBRARY

通过公开设计竞赛，DCM 赢得了印度尼西亚大学的中央图书馆项目。这项设计巧妙地结合了传统和创新。图书馆是一系列矗立于校园中、呈环形排列的塔楼，设计灵感来源于古印度尼西亚人在石碑上镌刻古训的传统——一组抽象的、刻有碑文的石碑式建筑屹立在绿草覆盖的圆形山丘上，将古代的传统延续到了现代。花岗岩覆层的石碑式塔楼高低不同，镶嵌有细长玻璃窗，使得光线可以透进下层空间。| 环形的建筑群经湖水侵蚀，形成一个环形剧场；郁郁葱葱的桃花心木掩映着湖光。图书馆的主入口设在环形的开口处，光线得以进入环形内部。在土丘和混凝土屋顶的下方，五层图书馆存放着珍稀的手稿、书籍和研究 / 参考书，这里常年恒温，且避免阳光直射。花岗岩挂板和细长的玻璃窗可以减少热能吸收，降低空调的负荷。

Selected in an open design competition, the scheme for a new university library deftly bridges the past and present. A series of towers projects from a circular landform. The concept takes inspiration from the ancient Indonesian practice of inscribing wisdom on stone tablets, and makes the leap into modern-day Indonesia as a series of abstracted stone tablets – prasasti – rising from the circular grass-covered earth mound. The granite-clad tablets of varying heights are 'inscribed' with narrow glazed bands, filtering light into the volumes below. | The circular landform is eroded on the lakeside, opening up to an amphitheatre with mature mahogany trees overlooking the lake. The opening identifies the main entry, and admits light into the interior. Beneath the mound's soil layer and concrete roof, five storeys of rare manuscripts, books and research/reference materials are housed in a stable temperature away from direct sunlight. Solid stone cladding and narrow bands of glazing further reduce heat gain, reducing the air conditioning load.

建设地点　印度尼西亚　雅加达
客　　户　印度尼西亚大学

Location　Jakarta, Indonesia
Client　University of Indonesia

怀特埃利沙
医学研究院
WALTER AND ELIZA
HALL INSTITUTE OF
MEDICAL RESEARCH

怀特艾利沙医学研究院是国际最尖端的医学研究中心之一。新楼的建设和旧楼的翻新使研究院的面积增加了一倍。| 新楼的外观设计成一组横向叠放的金属体块，为整个建筑群带来一种充满生机、积极向上的意象。每一个金属体块都采用缎面铝板外包，外立面镶嵌有玻璃，并设有细长的铝制遮阳百叶窗，窗口的开合角度不一。旧楼部分采用抛光铝板幕墙，与新楼的风格协调，外立面不规则的长方形开窗与新楼的百叶窗相呼应。| 新旧建筑之间由一个7层楼通高的中庭连通。人们可以通过中庭看到各个楼层，从观光电梯中可以看到实验室内部，这一空间设计营造出一种社区的氛围。| 新旧建筑包括了办公室、实验室、科学研究室、生物资源设施、一间养虫室以及先进的细胞组织成像和流式细胞计中心。会议设施包括了一个300座的讲堂。设计者为研究院创造出一个高品质的工作环境，为合作研究提供了空间，因而吸引了高水准的研究人员。

WEST ELEVATION

Walter and Eliza Hall Institute is at the forefront of international medical research. The new building and renovation of an existing building effectively double the Institute's size. | Carrying a vibrant, forward-looking imagery for the whole complex, the new building is conceived as a series of horizontally stacked metal boxes. Each is a satin silver envelope, brought to life with glazing shaded by banks of aluminium screening louvres with randomly placed downturned tab edges. Fusing it to the new, the older building is, in part, lightly wrapped in a new polished aluminium skin. Random rectangular openings echo the character of the louvred tabs on the new building. | The seven levels of each building link directly in a new atrium. The atrium offers visual connection between levels with views of the working laboratories from glazed lifts, enhancing a sense of community within the workspace. | The buildings house offices, laboratory and scientific research services, bio-resource facilities, an insectary, advanced cell and tissue imaging, and flow cytometry centres. Seminar facilities include a 300-seat lecture theatre. A high-quality work environment, fostering collaborative research is created to attract the highest calibre of staff.

new perforated plate
screens 'wrap' ends and
roof level of existing building
to establish new presence to
Royal Parade views + tie in
to new forms of WEHI 2.

existing bldg
level 9 reworked &
lower levels 'painted out'

WEHI 2 articulated as
3 finger elements
formed from 'wrapped plates'
with incorporated sunscreening.

Reimaging of WEHI from Royal Parade
to tie with new WEHI 2 expression.

建设地点	澳大利亚 墨尔本
客　　户	怀特艾利沙医学研究院

Location　Melbourne, Australia
Client　　Walter and Eliza Hall Institute of Medical Research

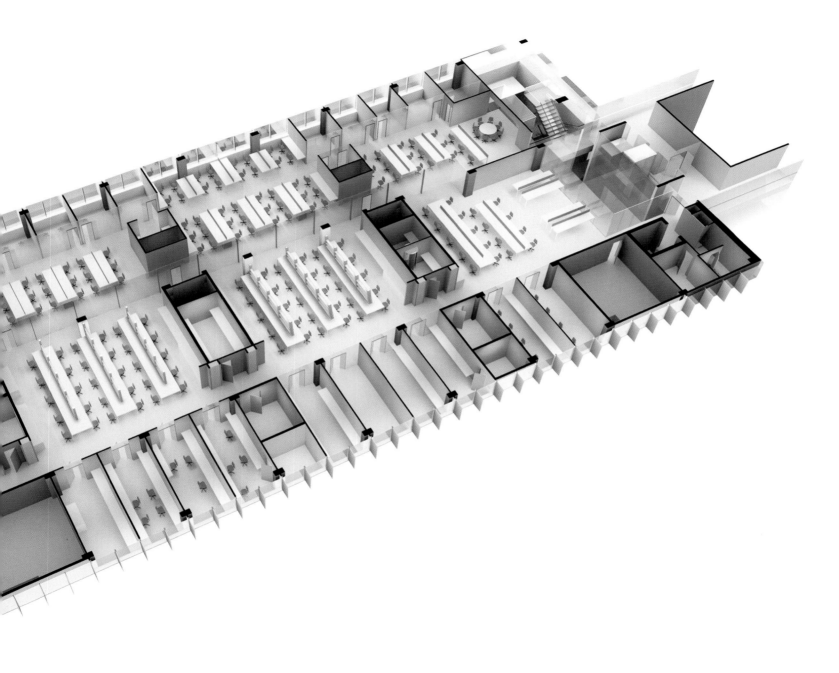

中国驻墨尔本
总领事馆
CHINESE CONSULATE
MELBOURNE

本项目为中国人民共和国驻墨尔本总领事馆的新馆。| 总领事馆新楼与原馆毗邻，设有地下车库，其中首层和二层是签证大厅和办公室，三层为外交官公寓。三层总建筑面积约为2500m²。

A new building for the Consulate-General of the People's Republic of China in Melbourne. | Built on a site adjacent to the existing Consulate, the building comprises basement parking, ground and first-floor visa offices, and four staff apartments on the second floor. The three-storey building comprises approximately 2500m² GFA.

建设地点 澳大利亚 墨尔本
客　户 中华人民共和国外交部

Location Melbourne, Australia
Client Ministry of Foreign Affairs, People's Republic of China

现代服务展示大厦
CONTEMPORARY TECHNOLOGY EXHIBITION BUILDING

现代服务展示大厦位于广东省东莞市松山湖畔，这是一个参加设计邀请赛的设计作品。| 展示大厦包括展示中心、信息中心和办公区。展示大厦的选址非常敏感，设计者希望在对环境的影响最小的同时，又能最大限度地利用基地特有的景观资源，为松山湖注入新的活力。最终，设计通过造型优雅的叠水景观，使松山湖环绕展示中心，将松山湖一直延伸到中心的入口处。入口广场宽敞，强化了绿色景观；从透明的入口大厅步入展示中心，人们可以一路欣赏美丽的湖景。从入口大厅可以通向均为4000m^2的展示中心和信息中心。这两个功能中心的外立面都采用了独立倾斜的玻璃和实体墙面。| 展示中心上方是1.3万m^2的办公空间。办公区的外立面采用经过艺术处理的金属外框，包裹着内部的竹林，而金属外框的设计灵感正是来源于竹林的意象。这个绿色环保外框将办公空间围绕在绿茵之中，可以有效地调节采光和通风，起到遮阳并净化空气的作用。| 办公楼层采用高效的线形空间布局，内部的采光天井种有绿色植物，使建筑可以自然通风。办公楼临湖的一面采用架空设计，外观十分独特。办公楼层架空的部分为其下方的机动车下客点、入口和建筑的四周遮蔽风雨和阳光。

An invited competition entry to design a contemporary technology exhibition centre on Songshan Lake in Dongguan, Guangdong Province. | The complex consists of Exhibition Centre, Information Centre and office space; and is carefully sited to minimise environmental impact and maximise the landscape and visual opportunities. The design celebrates the lake, extending it around towards the entry in an elegant series of stepped water terraces. A broad entry plaza provides a generous landscaped arrival point, with the transparent Grand Entry Hall to the Exhibition Centre allowing views through to the lake. Openings in the lower block reveal the entry halls leading to the 4000m^2 Exhibition Centre and the 4000m^2 Information Centre. Their facades are broken into angled glazed and solid panels. | Above, 13,000m^2 of office space is wrapped in a unique environmental veil – an exterior layer of stylised metal bamboo-inspired screening wrapped around an interior living bamboo garden. This surrounds the offices in a green wall, providing light, shade and filtered air. | The office floors are arranged in efficient rectilinear spaces with internal shaded green light courts that allow the building to breathe via natural ventilation spaces along its length. This floor is lifted at the lakeside to create a distinctive form. The overhanging office floors provide continuous weather protection to the vehicular drop arrival, building entries and entire building perimeter.

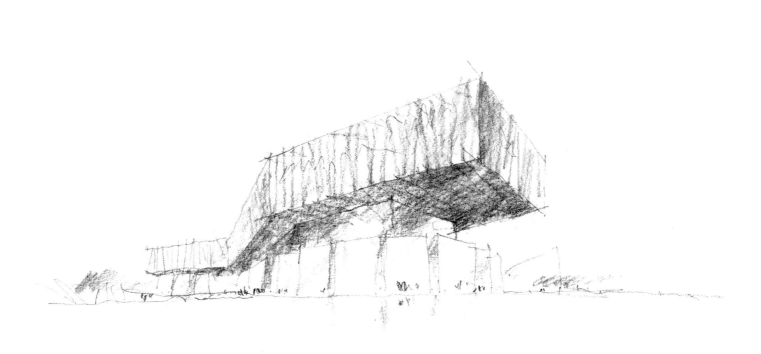

建设地点 中国 东莞 松山湖
客　　户 东莞松山政府

Location Songshan Lake, China
Client Dongguan Songshan Lake Local Government

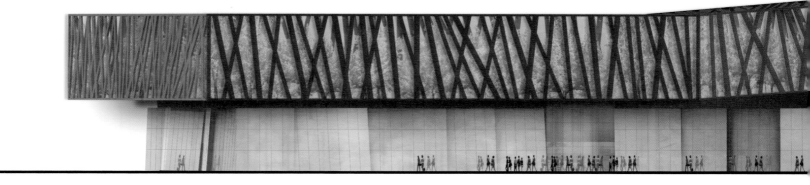

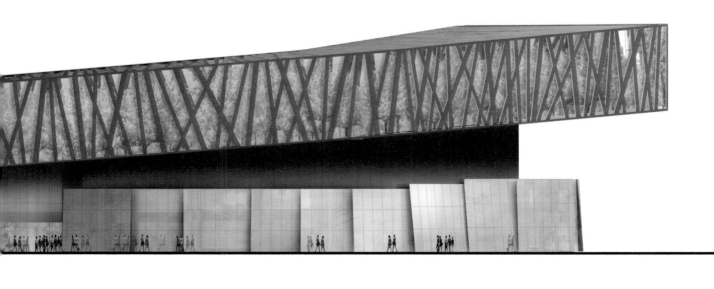

长沙滨江文化公园
CHANGSHA
RIVERFRONT
CULTURAL PARK

长沙滨江文化公园是为长沙滨江文化中心国际竞赛设计的作品。在外观上，抽象而成的巨龙造型蜿蜒贯穿整个公园。建筑表皮采用铝合金板，色彩抽象且富有金属光泽，其图案取材于传统中式花窗门的窗格。| 在整体漂浮的屋顶下，图书馆、音乐厅和博物馆相互独立，并拥有各自独立的入口。这些入口均设置在连续的屋顶下以便机动车下客。屋顶覆盖下的灰色空间同时可以作为户外文化活动的场所。| 图书馆中设有中庭，由中庭连接门厅和儿童图书馆；参考文献区和成人图书馆位于楼上。音乐厅的中心是一个1200座的主厅，另设有一个200座的小型音乐厅以及两个三四百座、可用于室内音乐或独奏的音乐厅。| 博物馆拥有八个主要展厅，面积在700m^{2}到1000m^{2}不等。博物馆沿河一侧是一个50m×40m，高25m的巨型"观河厅"，这里是欣赏公园和湘江美景的绝佳地点，并为各种特别活动、聚会庆典和展览提供了理想场所。| 酒店和会议中心位于毗邻的一幢60层高的标志性塔楼和裙楼中。

The dragon-like shape of this international design competition entry for a new cultural precinct for Changsha twists and turns through its park setting. The outer skin is composed of abstract coloured and metallic aluminium panels in a pattern that references traditional Chinese window glazing beads. | Contained within the overall twisting roof structure, the library, concert hall and museum are separate entities with clearly identifiable entrances, located under the continuous ribbon roof. Between these cultural buildings, the dragon roof provides covered plazas for drop-off, and acts as covered outdoor performance and gathering spaces. | The library is organised around a central atrium leading from the foyer to the children's library, with the reference section and adult library above. The 1200-seat hall, which forms the centrepiece of the concert hall, is accompanied by a 200-seat music hall and two 300-400-seat chamber music and recital halls. | The museum building contains eight major exhibition halls ranging in size from 700m² to 1000m². Adjacent to the museum, a vast 50m x 40m by 25m-high 'River Room' creates a spectacular space for special events, functions and exhibitions with its superb views over the park and Xiang River. | An adjacent, 60-storey urban marker tower includes a hotel and convention centre.

建设地点	中国 长沙
客　　户	长沙城市建设委员会
荣获奖项	国际设计邀请赛头奖

Location Changsha, China
Client Changsha City Committee of Construction
Awards First prize in limited international design competition

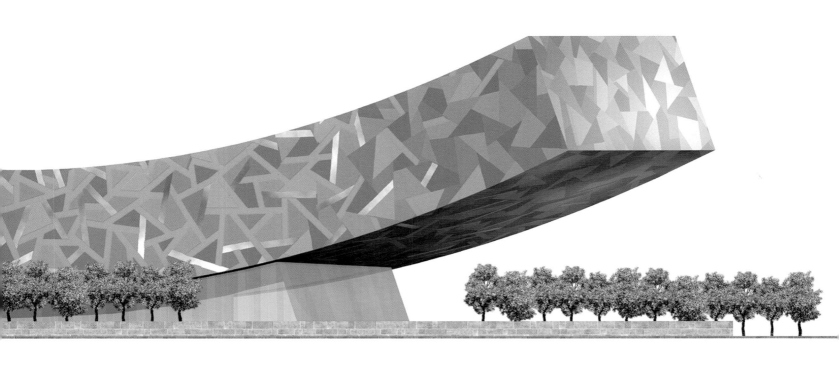

中国国家博物馆
设计竞赛
NATIONAL MUSEUM
OF CHINA
COMPETITION

这是DCM在中国国家博物馆的改扩建项目国际设计邀请赛中的竞赛作品。｜中国国家博物馆位于北京天安门广场上，无论是在历史上还是在今天，天安门广场都是北京乃至于中国的中心。因此，该项设计采取谨慎的手法，在扩建部分充分尊重原博物馆；新建部分的东馆设计为现代建筑，简约而大胆的外观既与原馆和谐共存，又展露卓尔不凡的气度。东侧的八座方形建筑采用了独特的嵌入图案，设计灵感来源于中国书法、建筑、城市规划和装饰艺术。｜室内的布局一目了然，直线型的中央大厅将一系列方形展厅串联在一起。｜DCM是六个进入决赛的国际建筑事务所之一。

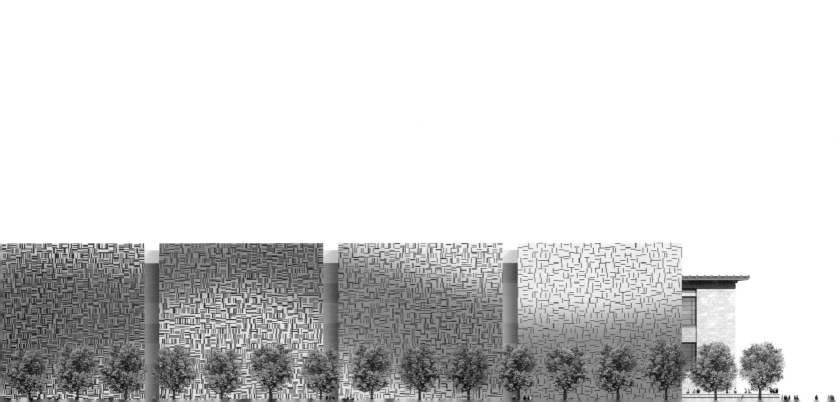

A limited international design competition entry for the redevelopment of the National Museum of China, situated prominently in Tiananmen Square, Beijing. | The design approach sensitively expands the existing museum in the urban context of Tiananmen Square, the epicentre of historical and modern Beijing. A new east wing forms a significant and memorable work of contemporary architecture with a simple yet bold form that exists in harmony with the existing museum, yet distinguished from it. The eight cubes of the east wing have a striking inlay pattern that resonates with Chinese calligraphy, architecture, town planning and decorative arts. | Internal planning is highly legible with the cubes housing a sequence of galleries accessed via a central concourse spine. | Denton Corker Marshall was one of six international architectural practices short-listed for the competition.

建设地点 中国 北京
客　　户 中国文化部

Location Beijing, China
Client Ministry of Culture, People's Republic of China

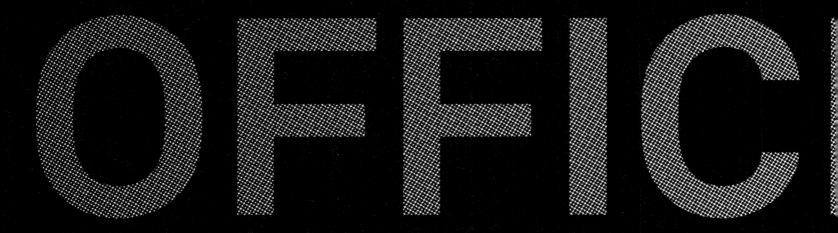

办公 + 商业

亚洲广场
ASIA SQUARE

亚洲广场坐落在新加坡滨海湾（Marina Bay）的新金融区，建筑面积 24.5 万 m^2，由高级办公塔楼和五星级威斯汀酒店组成。它是该地区最具环保效益的商业开发项目之一。｜双子塔从一座简约的正交形玻璃裙楼中跃然而出，两座塔楼均由八个细长且高度不一的柱形筒体组成。这一设计在新加坡新金融区的天际线上添加了鲜明的一笔。｜与塔楼相比，白色的点阵玻璃裙楼则造型简洁，形状完整。它正对着主街道，其设计理念是创造一个如同漂浮在海平面上的"冰块"。这个部分拥有 9000m^2 的公共空间，其中包括了人行空间、机动车下客点和一个重要的城市空间——城市客厅。它是城市公共空间的扩展，并且能够提供遮蔽，这样的设计旨在为各类活动，无论是非正式集会还是大型公共活动提供场地。｜亚洲广场的一号塔楼共有 43 层，其中 38 层（占总楼层面积的 94%）设置为专用的高级办公区，而一层和二层则设有商铺和餐饮设施。二号塔楼中有 26 层办公楼，在办公楼层之上则是占据 15 层高、拥有 305 间客房的五星级豪华酒店。两座塔楼的办公楼层中都设计了利用率高、无立柱的平面，标准层面积达 3000m^2~3490m^2。

Designed as one of the region's most environmentally efficient commercial developments, 245,000m^2 Asia Square comprises premium grade office towers and a five-star Westin Hotel in Singapore's new financial district at Marina Bay. | The towers emerge from a simple, orthogonal glass podium. Each tower is seemingly composed of a cluster of eight slender shafts that individually rise to varying heights to create a distinctive signature on the Singaporean skyline. | In contrast to the towers, the white fritted glass podium is expressed as a singular, unifying element. It formally fronts the primary streets and is conceived as an 'ice block' floating above the ground plane. It contains a 9000m^2 public space, including covered pedestrian footpaths, vehicle drop-off and a major urban space — the City Room. This is a sheltered extension of the urban public realm designed to accommodate anything from casual meetings to major public events. | Asia Square Tower 1 is a 43-storey office building with 38 floors (94 per cent of floor area) devoted to A-Grade office space. The first and second floors contain a mix of retail space and food and beverage outlets. Its companion, Asia Square Tower 2, comprises 26 office levels surmounted by a 15-level, five-star luxury hotel of 305 rooms. Each provides a highly efficient, column-free floor plate ranging from 3000m^2 to 3490m^2 GFA.

建设地点	新加坡
客　　户	MGPA
荣获奖项	2010 年美国绿色建筑协会颁发的 领先能源与环境设计结构外观铂金预发证书奖 2009 年新加坡建设局绿标铂金奖

Location	Singapore
Client	MGPA
Awards	- Awarded the Leadership in Energy and Environmental Design Core and Shell (LEED-CS) Platinum pre-certification by the US Green Building Council, 2010 - BCA Green Mark Platinum Award, 2009

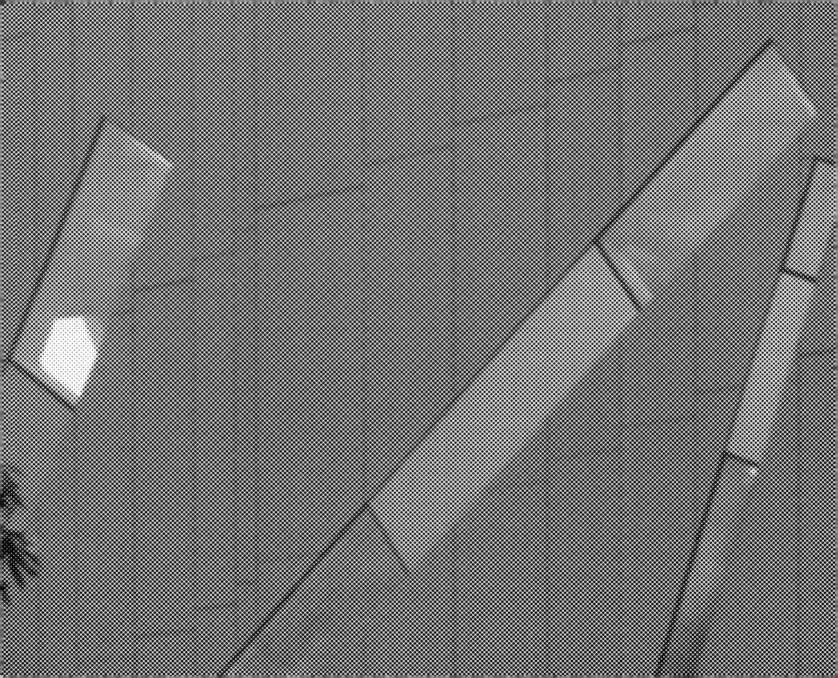

布里斯班广场
BRISBANE SQUARE

布里斯班广场是一座位于布里斯班市重要地段的地标式建筑,西临布里斯班河,毗邻繁忙的北港区,并直接通往皇后大街步行街。| 办公塔楼加裙楼共 37 层高,总建筑面积为 6.5 万 m²。布里斯班市政厅的办公室和公共图书馆皆设在此,同时还安排了商业设施、公共广场以及业主办公空间。| 作为布里斯班市一个重要的社交和文化中心,建筑首层专用于城市公共活动。商铺、咖啡厅和餐饮场所设置在街道层面,并与邻近财政大楼前方的公共区域呼应。四个低层柱形空间容纳了图书馆、政府市民接待中心和其他办公区域,丰富着广场的色彩。| 地面 6 层以上是塔楼,具有两种特征鲜明的构件:黄色的柱子划分出顶楼 10 层供业主使用,玻璃幕墙外立面上的穿孔遮阳板划分出市政厅办公室的界限;遮阳板的设置提高了高效玻璃幕墙的环保效能,体现出市政厅对生态可持续发展原则的重视,同时又赋予市政办公室鲜明的造型特点。| 布里斯班广场是通往市中心的一个地标建筑,它为市区天际线增添了绚丽的一笔。

Brisbane Square is a landmark building located in a key urban position, fronting the Brisbane River and North Quay, and with direct links to Queen Street pedestrian mall. | The 37-storey office tower and podium of 65,000m² accommodates Brisbane City Council's offices and public library, with retail facilities and a public plaza, as well as commercial office space for the building owner. | Designed as an important social and cultural hub for the city, the ground plane is dedicated to the public domain. Shops, cafes and food outlets at street level open onto a public space in front of the neighbouring old Treasury Building. Floating over the plaza and providing splashes of colour are four low-rise linear stick buildings housing the council library, customer-service centre and other office space. | The tower sits six storeys above ground and has two distinct sections. Yellow sticks set apart the top 10 levels for use by the owner, while a perforated sunscreen over the glass façade demarcates the offices of the city council. The sunscreen speaks to the environmental efficiency of the high-performance glazed curtain wall, illustrating the council's commitment to ESD principles and providing a distinct identity for the city offices. | Brisbane Square is an emblematic gateway to the city and a distinctive addition to the city's skyline.

建设地点 澳大利亚 布里斯班
客　　户 布里斯班市政厅
　　　　　Suncorp Metway 银行投资管理有限公司
　　　　　荷兰银行

Location Brisbane, Australia
Client Brisbane City Council
Suncorp Metway Investment Management Ltd
ABN Amro

乐天银泰百货
INTIME LOTTE
SHOPPING CENTRE

乐天银泰百货位于吉祥京剧院的原址，地处北京购物黄金地段，是一座集购物、娱乐和剧院于一体的综合开发项目，面积达到 5 万 m²。| 从当地历史渊源和现代使用功能出发，设计者在建筑外观上采用了一道晶莹玻璃帷幕，而将主建筑设计为剧院的舞台。帷幕后方的室内色彩纷呈、富有动感，并设有广告和展示空间，促发与往来于王府井大街和金鱼胡同的行人之间的对话。| 有别于传统商业建筑中运用橱窗展示和标志的手法，乐天银泰百货的建筑本身就是一个巨大的广告。它是一个富有现代感，充满生气与活力的作品，将王府井整条街的生活万象带到它的舞台中央上演。| 波浪形的悬挂式幕墙形成一个透明程度不一、反射和折射效果变化无穷的界面，将来往行人的影像投射到建筑表面，形成梦幻般的动态效果，丰富的变化一改传统建筑静止、被动的形象。| 这个建筑凭借国际商业设计前沿的设计概念，在一个富有活力的城市中心创建了一个地标式商业中心。

On the site of the original Ji Xiang Peking Opera Theatre, the 50,000m² shopping, entertainment and theatre complex is located on Beijing's premier shopping street. | In response to its historical artistic context and contemporary program of uses, the building is conceived as a theatre behind a transparent glass stage curtain. Behind this curtain, an interior of colour, movement, advertising and display is revealed to the Wang Fu Jing and Jin Yu Hu Tong streets. | Instead of a 'traditional building' with windows and signage, here the building becomes one huge sign; contemporary, vibrant, exciting, and memorable. The life and activity of the street is drawn upwards and into its heart. | The sensuous draped perimeter skin provides an ever-changing surface of reflection and transparency, where the movement of passers-by creates an illusion of movement and complexity in an otherwise calm and understated form. | The design concept is at the forefront of international retail design and creates a landmark shopping centre in the heart of a confident city.

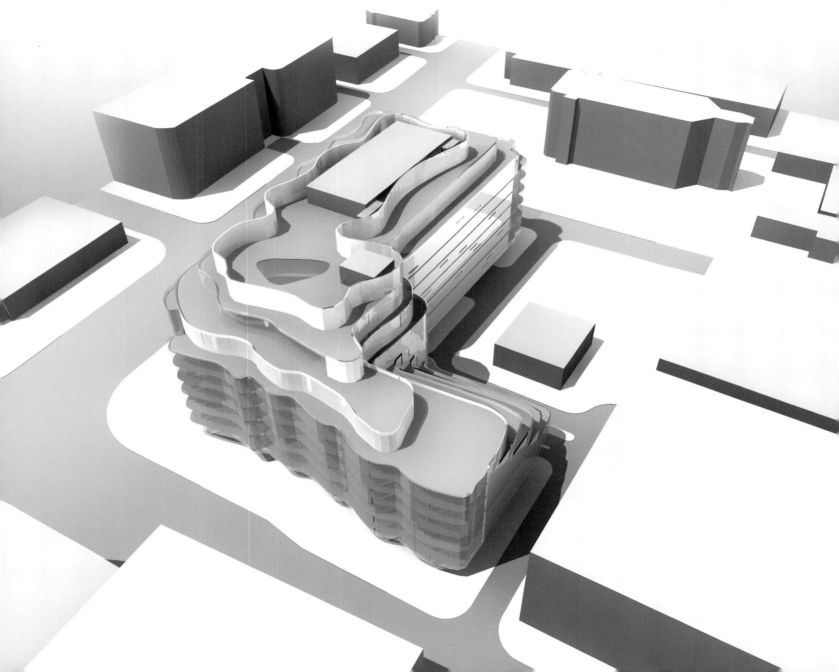

建设地点 中国 北京
客　　户 中国 银泰

Location Beijing, China
Client Yintai Investment Company

森斯办公楼
SENSIS
BUILDING

森斯办公楼共五层,建筑面积 2 万 m²,位于墨尔本市中心维多利亚女王综合商业裙房之上。森斯因其简约优雅的线形外观而显得格外出众,同时与周边略为嘈杂的商业中心形成鲜明的对比,愈发显得宁静。| 森斯大楼首层设有门厅。一对黄色的斜柱成为建筑入口的标志,上方为一组上下错落的不同材质的盒子。| 森斯大楼是墨尔本城市天际线中颇有特点的组成部分,从周围甚至较远距离的建筑内都可以观赏到它。设备用房设计为色彩鲜明的长柱,以看似随意的布局横放在屋顶上,这一设计使建筑在周围的环境中更脱颖而出。

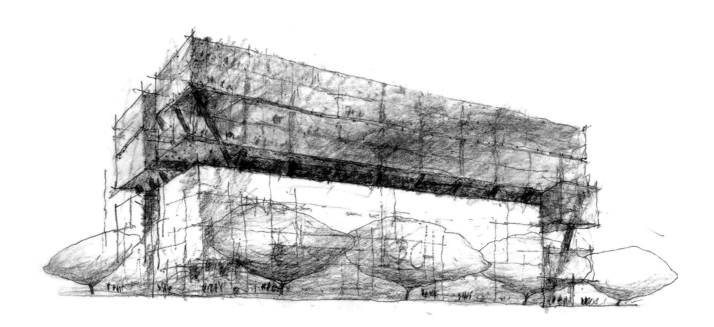

A five-storey commercial office building of 20,000m² that sits above the QV Village retail complex in central Melbourne. The design's simple, elegantly detailed rectilinear forms assist in delivering a separate identity. The Sensis building floats in strong contrast to the somewhat chaotic forms of the QV complex buildings below. | The building establishes its address via a ground level foyer. The foyer is visually connected to the floors above by distinctive inclined columns supporting a complementary transparent box form. | As a building that resides in an urban roofscape, it is widely seen at relatively long distances from surrounding city buildings. Its presence in this environment is emphasised by configuring the plant rooms as large colourful stick forms placed, seemingly casually, on the roof.

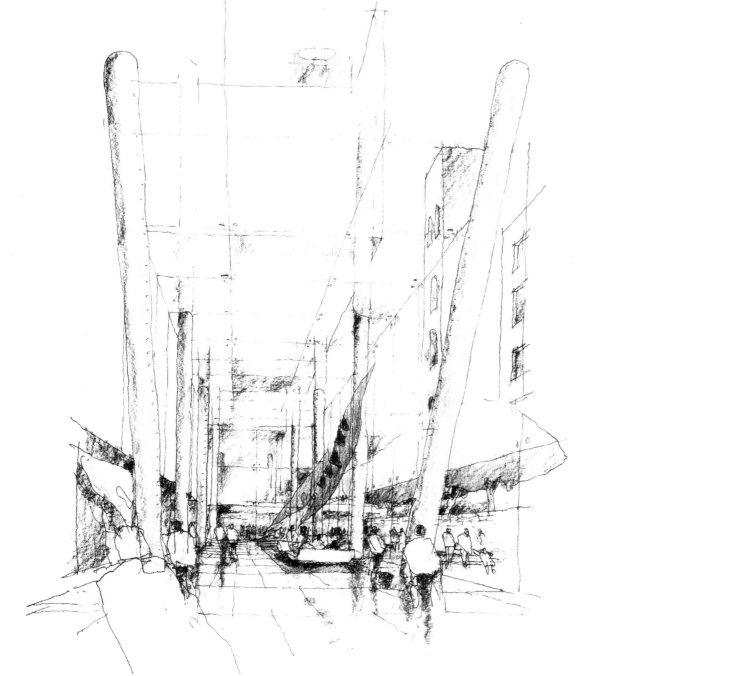

建设地点 澳大利亚 墨尔本
客　　户 Grocon 有限公司

Location Melbourne, Australia
Client Grocon Pty Ltd

中坤大厦
ZHONGKUN TOWER

中坤大厦是一座具有雕塑般外形的、19层高的综合办公楼，是城市文脉中一座特征鲜明的地标式建筑。中坤大厦的建筑面积达7万 m^2，是一个拥有15hm^{2}规模的住宅、商务、商业和休闲重建项目的一部分。| 大厦坐落在北京动物园附近的南长河畔，设计通过将建筑形体分解成片状结构而改变视觉体量。外立面上设有横向的深浅不一的带状玻璃窗，以及钛合金涂层的的铝挂板，二者交错有致，形成独特的外观效果。巨型透明玻璃如同一个"城市窗口"，进一步巧妙地改变了建筑的原有尺度。

The sculptural form of this 19-storey office complex creates a distinctive landmark in its urban setting. The 70,000m² Zhongkun Tower forms part of a 15-hectare redevelopment of residential, commercial, retail and recreational space. | Set on the Nanchang River near the Beijing Zoo, the building's visual bulk is reduced by breaking the form into finger elements. The distinctive striped appearance of the facade is created with horizontal window panels of dark and silver glass interspersed with strips of titanium-paint finished aluminium. Large clear glass 'urban windows' further play with the apparent scale.

298 OFFICES + RETAIL

建设地点　中国　北京
客　　户　中坤锦绣房地产开发有限公司

Location Beijing, China
Client Zhongkun Jinxiu Real Estate Development Corporation

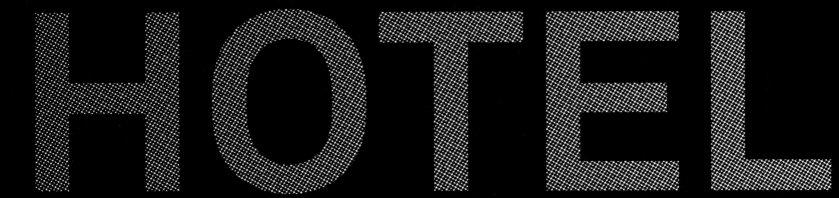
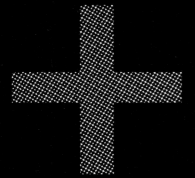
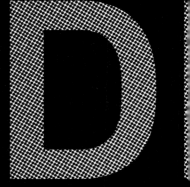

RESI-
ENTIAL

丽山别墅
与酒庄
VIEW HILL
HOUSE
AND VINEYARD

丽山别墅坐落在 6h 的小山顶上，俯瞰 32hm² 的葡萄园，将整个亚拉河谷的美景尽收眼底。| 别墅由两个金属筒组成，如同一件独特的雕塑作品。其中，一条耐候钢筒横在地面，上方的另一个金属筒采用无光暗黑色铝材。上下两个金属筒相互垂直，上方的金属筒前后分别向外悬挑 6m 和 9m。两个金属筒精确地朝向南北和东西。从上空俯瞰，别墅位于一个澳大利亚测量基准点标志的边上，如同一个不对称的十字放在罗盘的方位基点上。| 室内设计与外观一样，采用极简美学的设计手法。室内面层采用灰绿色刨花板饰面。卧室、两端的办公室和中间的起居室之间设有白色的隔断。可开启的穿孔板将别墅侧面的双层玻璃窗遮蔽起来。金属筒北侧的外墙可以向上开启三个不同角度，为长达 10m 的起居空间提供遮阳。

Set on top of a hill the house overlooks a 32-hectare vineyard set on a 6-hectare property, with views across the entire Yarra Valley. | The simple proposition of two long metal tubes presents as a striking piece of sculpture. A Corten tube rests on the ground; another, in matt black aluminium, sits somewhat precariously on top at right angles, cantilevering 6m and 9m at the front and back. The tubes are precisely aligned north-south and east-west. Viewed from above, the house – sited next to a trigonometric point – is an asymmetrical cross marking the cardinal points of a compass. | The inside is formed by the minimalist aesthetic that shapes the exterior. A second skin of grey-green stained strand board lines the interiors. White boxes break the internal volume into bedrooms and offices on the ends and living space in the middle. Double glazed windows in the sides and the long metal tube are concealed behind perforated panels. A large north-facing section of the Corten box lifts up in three sections, creating sun shade across the 10m wide living area glazing.

建设地点 澳大利亚 维多利亚州 亚拉河谷
客　　户 约翰·丹顿

Location Yarra Valley, Victoria, Australia
Client John Denton

梅德赫斯
酒庄别墅
MEDHURST
HOUSE

梅德赫斯酒庄别墅坐落在墨尔本东北部的亚拉河谷。它位于一个小山丘上，俯瞰北侧的葡萄庄园。｜别墅的楼板和屋顶采用两片长方形薄板，缓缓升高到葡萄园之上，向主干道方向延伸。黑色的金属楼板由一组与之垂直的黑色混凝土厚墙支撑。楼板一端的悬臂结构较短，而另一端则向外出挑达 11m。屋顶盖板采用与楼板相同的构件，其间是通高玻璃窗。在别墅背部的立面上，楼板和屋顶之间镶嵌有两个绿色横条。别墅内部为钢结构，前后 2m 的缩进空间和两端 5m 的悬臂使两片薄板显得更为独特简约。｜一层空间置于黑色混凝土厚墙之间，包括入口、车库、酒窖、书房和客房。二层采用简约的设计概念，是一个完整的一体化空间，包括主卧室、起居室、餐厅和厨房。内部空间由悬铃木护墙板分隔，护墙板与天花板之间用玻璃墙面分开。

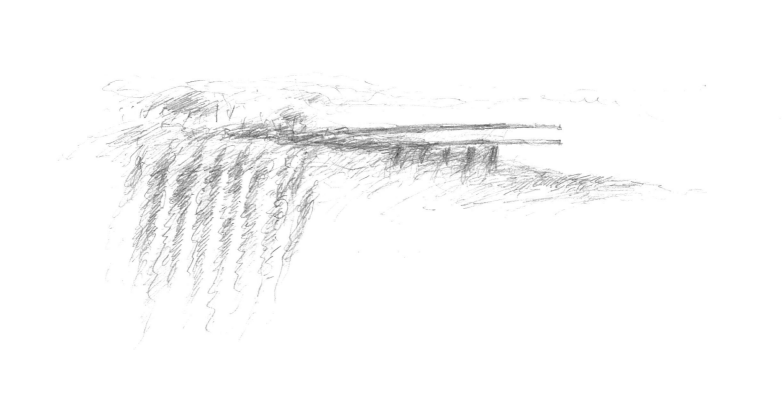

A house set above a vineyard in the Yarra Valley, north-east of Melbourne. | Two thin rectangular plates – roof and floor – lay in a gentle rise above the vineyard stretching from the main road. The black metal floor plate is supported by a series of parallel black pigmented concrete walls set at right angles to it. One end cantilevers gently and low; the other cantilevers 11m beyond the supporting wall. The identical roof plate floats above, separated by full height glazing on the front and ends. On the back two green sticks are laid longitudinally between the plates. Internal steel columns support the roof plate. The singularity and clarity of each plate is reinforced by deep setbacks – 2m on the front and rear, and 5m on the cantilevered projection – to the external wall. | Defined by the black concrete walls, the lower lever contains entry, car parking, wine cellar, study and guest bedrooms. The upper level is conceptually a single space and contains bedrooms, living, dining, and kitchen. Internal spaces are defined by sycamore panelled blocks inserted into the space, and held free of the ceiling and the long glazed wall.

建设地点 澳大利亚 维多利亚州 亚拉河谷
荣获奖项 2008 年澳大利亚联邦建筑师协会（维多利亚州）住宅建筑奖
2008 年澳大利亚联邦建筑师协会住宅建筑国家表彰奖

Location Yarra Valley, Victoria, Australia
Award - Award for Residential Architecture,
 Australian Institute of Architects (VIC), 2008
 - National Commendation for Residential Architecture,
 Australian Institute of Architects, 2008

落樱酒庄
ZHONGKUN
VINEYARD

落樱酒庄坐落在北京北郊，临近长城，是一项高端的豪华酒庄开发项目。| 项目占地100hm^2，在田园风光的环绕中，依傍绵延的群山。游人可沿一条800m长的林荫大道进入酒庄，体验宾至如归的氛围。| 绵延的石墙上是几条金属长筒，金属管自由地斜跨在石墙之上，设计极富表现力和动感。建筑的造型配合地形，层叠的石墙与自然景色融为一体。| 落樱酒庄包括一座先进的酿酒厂、酒窖、品酒设施和一座拥有60间套房的小型精品酒店；另外还包括一间艺术画廊、高级餐厅、水疗中心、会议中心和一个露天音乐厅。项目开发的第一期还包括33套豪华会所——每座会所的面积都达1500m^{2}以上。

Located north of Beijing near the Great Wall of China, this exclusive vineyard resort development is high-end in all respects. | The 100-hectare site sits in rural land sheltered by a mountain range. A gradual approach along an 800m avenue heightens the sense of arrival. | The expressive gesture is of long walls with sticks on top, and a skewed metal tube resting jauntily on the sticks. The buildings are configured to follow the land contours. Submerged into the landscape, they come into view as a succession of stacked stone walls. | Zhongkun Vineyard comprises a state-of-the-art winery, cellars, wine-tasting facilities and a six-star, 60-suite boutique hotel; it also includes an art gallery, premium quality restaurants, a day spa, conference venue and an amphitheatre. The first stage of the development also includes 33 luxury clubhouse villas each with a generous floor plan of around 1500m^2.

建设地点 中国 北京 延庆县
客　　户 北京中坤投资集团

Location Yanqing County, China
Client Beijing Zhongkun Investment Group

欧容路 590 号
590 ORRONG ROAD

这是一个对内城区的旧城进行改造的整体规划和建筑设计项目。设计核心是一个小区广场，广场直通毗邻的轻轨车站。广场营造出一种宾至如归的氛围，并与周围的城市空间紧密结合。一条线形中央绿化带贯穿整个项目地块，为居民和社区提供了极为舒适的自然环境。| 建筑群由一组化零为整的小体量建筑单体组成，建筑高度从广场周边最高的十二层逐渐降低到基地边缘的六层；两层和三层联排别墅则对街道景观进行了重构。建筑的形体塑造在满足楼层面积需求的同时，为居民提供通透的视野，让人们可以欣赏到公园的风景和天空映衬下的城市轮廓线，而建筑群的布置还避免影响到项目毗邻的公共绿地的日照要求。| 该项目包括 466 套不同类型的住宅，一间便利店，拥有 130 个座位的快餐咖啡店和妇幼保健中心。设计者还设计了一个集公共活动室、泳池、健身房等多种功能的设施供小区的居民使用，并安排了 686 个停车位和 688 个自行车位。

Masterplan and architectural design for a residential development. At the heart of the design is a public plaza adjacent to a suburban railway station. It creates a strong sense of arrival and strengthens the connections to the surrounding urban fabric. From the plaza a linear central green leads up through the site creating a high degree of amenity for the residents and the community. | The built form is expressed as a series of small volumes with a maximum height of 12 storeys around the plaza, graduating down to six levels towards the perimeter of the site with two and three level townhouses completing a streetscape interface. The built form is located to achieve specific floor areas, views to parkland and the city skyline, visual permeability of the site and minimal overshadowing of adjacent parklands. | The site includes a total of 466 dwellings, a convenience shop, 130-seat licensed café, and maternal health centre. The design also proposed an ancillary multi-purpose room, pool, and gym facilities for use by the on-site residents, and a total of 686 car spaces and 688 bicycle spaces.

建设地点　澳大利亚　墨尔本
客　　户　Lend Lease Development Pty Ltd

Location　Melbourne, Australia
Client　Lend Lease Development Pty Ltd

欧景城市广场坐落于南宁市中心的林荫大道上，是一项大规模的城市重建开发项目。设计者采用了简约、巧妙且颇有趣致的几何结构，将建筑面积多达 21 万 m² 的住宅、商业、办公、酒店和休闲设施整合在一起。鲜明的色彩和不规则的图案赋予了建筑群生气与活力，使几何形式的外观独具表现力。| 项目中的八座塔楼高度不一，其中最高的一幢达 25 层，下方坐着 5 层裙楼，包含了两层商业区和三层 SOHO 单元。临街的彩色柱廊连成一条直线，建筑紧靠街道，营造出强烈的城市氛围。| 弧线型的住宅板楼位于临街建筑的后方，另有一组住宅板楼位于基地的北面，高度在 6 层到 8 层之间。这些住宅楼围合形成中央城市广场，面宽不一的明黄色阳台使建筑外立面饶富生趣。为适应亚热带气候特点，临街的住宅塔楼设有开敞的内廊，这一独特的设计使塔楼内部有良好的通风。| 沿街塔楼采用了一个外包框架，将复杂的建筑外形整合在一个统一的建筑元素之中。而随机扩大的阳台为每个住户创造出各自带有遮阳的室外活动空间。同时，遮阳板的设计也为建筑带来了生动的变化。开发项目中还包括 1.5 万 m² 的高级商业区，800 套公寓，1000 套 SOHO 单元和 650 个车位。

Located on Nanning's principal boulevard, this large urban renewal development employs simple, smart and playful geometry to unite 210,000m² of residential, retail, commercial, hotel and recreational facilities. Bright colour and random patterns enliven the composition, incorporating expressive qualities into the formal geometry. | Continuous five-storey podium buildings, comprising two levels of retail and three levels of SOHO units, sit beneath eight towers ranging in height up to 25 storeys. Ordered with a giant colonnade of brightly coloured columns, the buildings have minimal setbacks from the street to create a strong urban presence. | Curved residential towers are set behind the street-front buildings with six to eight-storey residential slab block buildings beyond to the north. The curved buildings act as a container to form an open urban square. Balconies in bright yellow of varying lengths playfully animate the facades. In response to the subtropical climatic conditions, the residential towers along the street feature distinctive breezeways that course through the buildings. | Towers along the street are wrapped in an external frame structure, unifying the complex building form into one single unified architectural element. Randomly enlarged balconies create shaded outdoor activity space for residents and the louvre screen panels bring more life to the building. | The development includes 15,000m² of premium retail space, 800 residential apartments, 1,000 SOHO units and parking for 650 cars.

建设地点 中国 南宁
客　　户 北京银信光华房地产开发有限公司
荣获奖项 2002 年中国建设部杰出设计奖

Location Nanning, China
Client Beijing YinXin GuangHua Real Estate Development Ltd
Awards Ministry of Construction Excellence in Design Award, 2002

杰克逊港
旧区改建
DISTILLERY HILL AT
JACKSONS LANDING

杰克逊港的蒂斯提山住宅开发项目位于悉尼的派蒙区，面朝悉尼海港大桥，包括两幢塔楼和周边的裙房。其中，蒂斯提塔楼高19层，包括95套公寓单元；考利塔高18层，包括89套公寓单元；另外还有沿街的低层裙房，包括17套公寓单元和12套联排别墅。项目设有3层地下停车场，住宅周围有景观花园、游泳池和健身房，周围海港美景尽收眼底。| 两幢塔楼采用简洁几何造型，使建筑从远处亦易于辨认。设计者在简化建筑形式语言以提升整体清晰度的同时，亦采用了随机的的百叶遮阳板（固定和活动式）来体现鲜明的特征。独特的彩色墙体比建筑顶部高出三层，使塔楼如同一尊雕塑矗立在地面上。| 蒂斯提山建筑群的高度和色彩，为悉尼天际线增添了亮丽的一笔。

A residential development of two towers with adjacent terrace housing in Pyrmont, looking out to the Sydney Harbour Bridge. | The 19-storey Distillery Tower comprises 95 apartments, the 18-storey Quarry Tower comprises 89 apartments, and a further 17 apartments and 12 townhouses in low-rise terraces activate the street frontages. With three levels of basement car parks, the development is set in landscaped gardens with a podium level pool/gymnasium located to capture harbour views. | The built form of each tower is a simple clear geometric composition – important for the reading of the tower from a distance. While reducing the built form to enhance overall clarity, the individuality of the apartments is reflected in the random pattern of fixed and moveable louvre screen panels. Coloured wall plates extend to street level, anchoring the tower firmly to the ground in a distinctive sculptural gesture. | Combining colour with height, the architecture makes for a distinctive addition to the Sydney skyline.

372 HOTEL + RESIDENTIAL

建设地点 澳大利亚 悉尼
客　户 Lend Lease Development Pty Ltd

Location Sydney, Australia
Client Lend Lease Development Pty Ltd

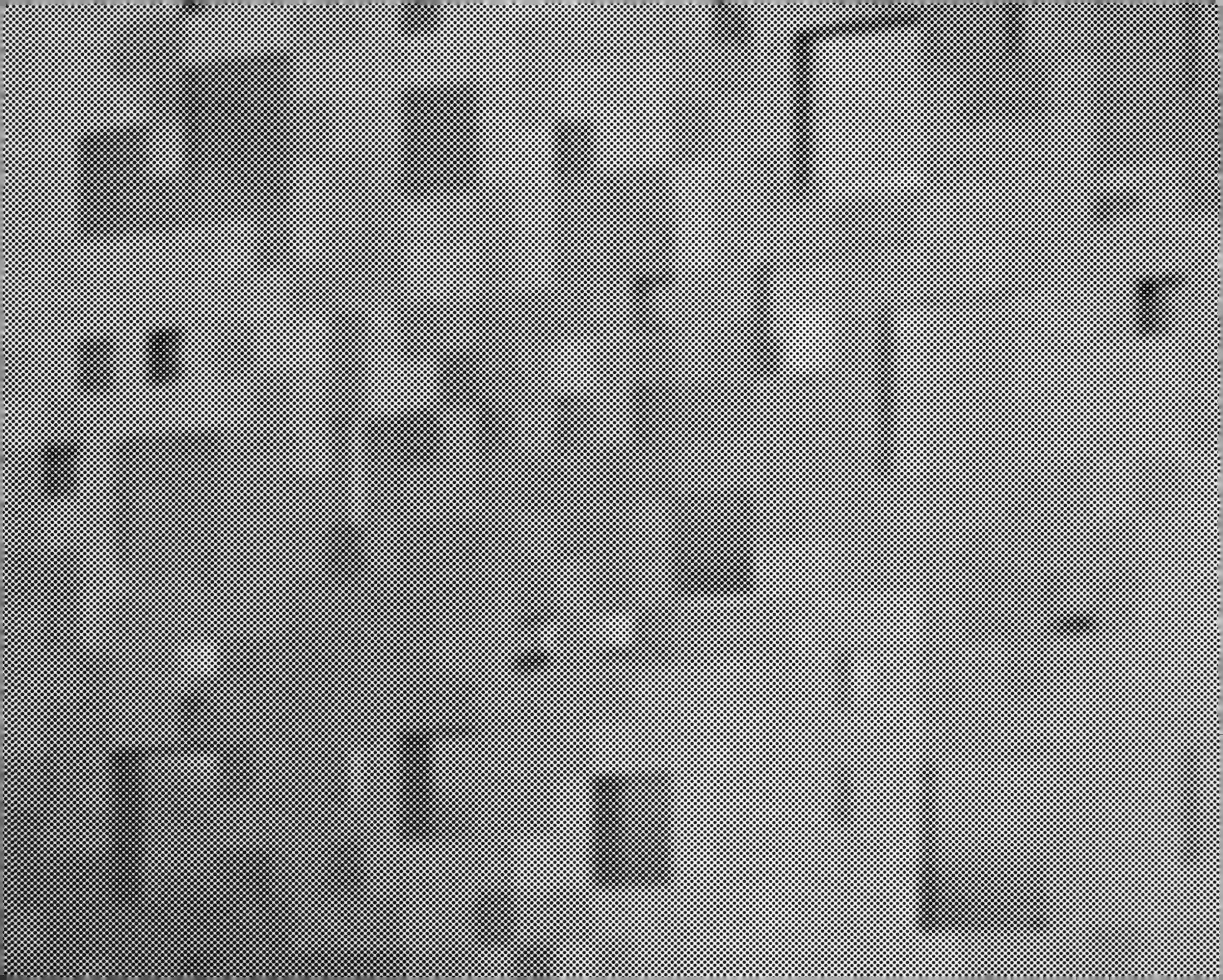

温莎酒店
THE
HOTEL WINDSOR

温莎酒店是澳大利亚历史上唯一的大酒店（Grand Hotel）。在对这个列入文物保护名单的历史酒店进行改扩建的项目中，设计者既保留了建筑固有的历史韵味，又使其富有焕然一新的现代气息。改建后的酒店由三个建筑体块构成，旨在通过这种设计恢复其往昔的风采。| 在改扩建工程中，建筑外观是对原温莎酒店的忠实复原，尤其是在沿街一侧重新引入原先首层中的面街廊柱，通过这样的设计还原出酒店历史韵味来。| 新建部分中，位于街角的建筑采用了简洁的多孔方块结构。开口布局看似随意，实则是酒店客房大小不一的窗户，这种设计为建筑外观带来动态的视觉质感。毗邻的原温莎酒店具有19世纪的建筑风格，而这座新楼则引发了这种建筑风格在21世纪的新的表现形式。| 另一幢新建建筑是坐落于温莎酒店后方的细长板楼，外观既具现代特色又低调含蓄。点阵玻璃立面营造了一种涟漪的效果，如窗纱一般成为温莎酒店的背景，使温莎酒店的古老塔楼看起来格外分明。这幢新的酒店客房大楼距离老楼外侧25m，从建筑体块的角度来说，这一后退距离足以使其看上去就像和温莎酒店毫无关联；然而，和周边其他文物建筑与新建筑的布局关系相比，这种距离的设置看上去恰如其分。| 该项目为酒店增加了152个客房和套间，使酒店客房总数增至332间，此外还新建了会议设施和拥有一个25m游泳池的健身休闲中心。由此，温莎酒店达到了现代五星级酒店标准。

Redevelopment of a heritage-listed hotel combines the existing historic fabric with a fresh contemporary expansion. | The tripartite composition is designed to return the Hotel Windsor to its rightful place of pre-eminence as Australia's only Grand Hotel. | The original Hotel Windsor is restored faithfully to its original exterior appearance – in particular, the re-opening of the colonnaded ground-level frontage. | A new corner building is to be a simple perforated cube. The seemingly random perforations provide varying sized windows to the hotel rooms within, their modulation and visual texture invoking a 21st century reflection of the adjoining 19th century frontage. | An ultra-slim tower with a rippling façade of seraphic glass rises up behind, forming a backdrop – literally a curtain – that allows a clear reading of the profile of the Windsor's historic towers. Set back 25m from the façade of the heritage building, the external expression of the new guestrooms is contemporary yet restrained. It is set far enough back to be read as a building unconnected with the Hotel Windsor itself, an arrangement that compares very favourably to other heritage/new built forms in the surrounding streetscapes. | Adding 152 rooms and suites to increase the total number to 332, in addition to new meeting facilities and a health and leisure centre with a 25m swimming pool, the redevelopment will bring the Hotel Windsor up to contemporary five-star standards.

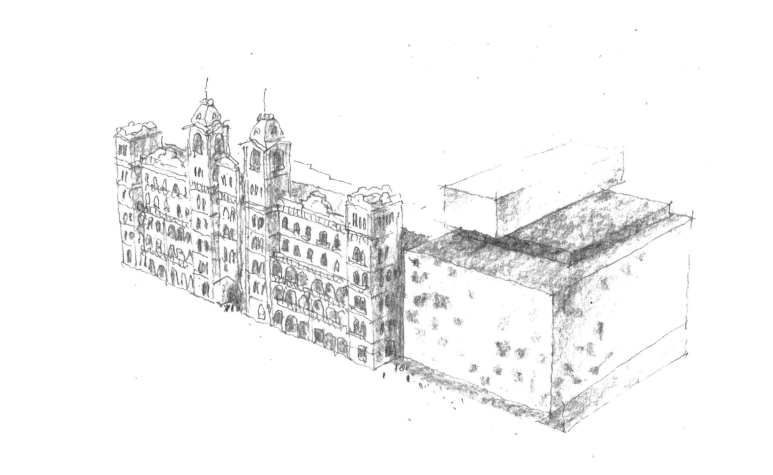

建设地点 澳大利亚 墨尔本
客　　户 Halim 集团

Location Melbourne, Australia
Client Halim Group

柳州城市广场 · 丽笙酒店
LIUZHOU CITY PLAZA
AND RADISSON BLU HOTEL

此项目旨在为柳州市新区创建一个大型城市综合社区，体现了多元化的设计理念。柳州城市广场毗邻新市政府行政中心和中央公园，它包括了各种户型的公寓塔楼、多层住宅、大型商业区、办公设施、SOHO 娱乐休闲设施和五星级的丽笙酒店。| 广场北面的办公楼采用类金属板和玻璃作为外立面的材料。规模较小的板楼、SOHO 塔楼和裙楼根据建筑形式和建筑类型进行分层，使公共空间和私人空间融为一体。每一幢公寓塔楼的处理方式各有不同，可以越过广场观赏到远处的山丘。在西面的入口广场的景观设计上，用林荫覆盖广场，从这里可以通往办公区、酒店和公寓。错落有致的公园小径、相连的池塘和其他景观元素，再加上现有的公园，共同创造出完美的空间布局和一个具有现代气息的居住环境。| 每一幢建筑皆采用精致的玻璃、混凝土和金属包层；建筑内设有空中的景观花园，采用遮板以增加进深、调节气候以提高环境舒适度。在整个广场范围内，精心调配的色彩成为亮点，它同时还是控制建筑布局和规模的一种方式。| 五星级的丽笙酒店坐落在柳州城市广场一角，包括 260 间客房，建筑面积 2.5 万 m^2，包括一个可容纳 300 人的宴会厅，几间不同风格的餐厅和设有泳池和健身房的休闲中心。酒店外观被设计成两片薄板，外立面采用金属孔板的设计旨在消解立面的尺度，创造一种轻盈的视觉效果。东西两侧的玻璃幕墙使酒店可以获取更为清晰的视野。

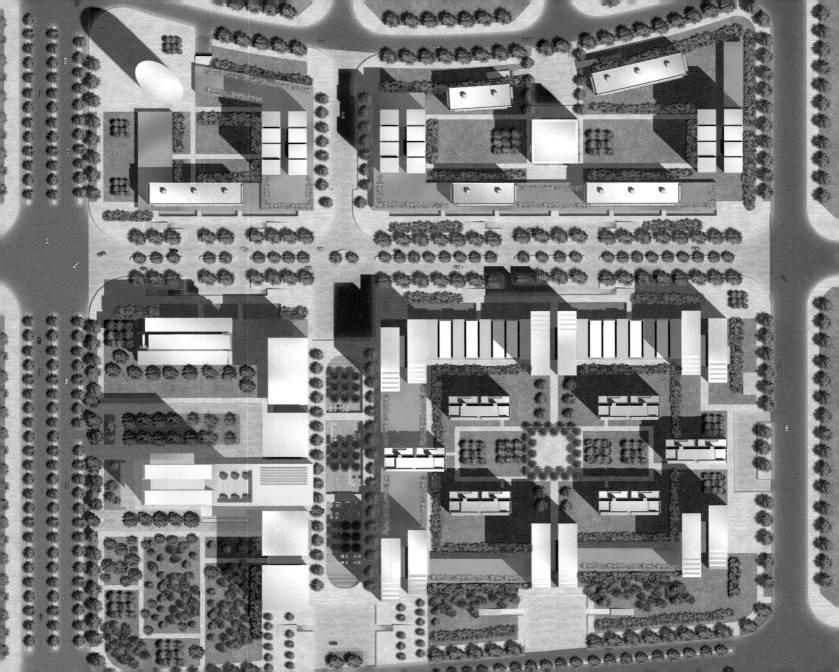

Coherent design vision delivers a major integrated urban neighbourhood for the city of Liuzhou in China. Located adjacent to the new government administrative centre and central parklands, the design provides a varied mix of apartment towers and lower rise housing mixed with major retail, office, SOHO recreational facilities and the five-star Radisson Blu Hotel. | To the north of the site, an office tower is sheathed in glass and metal panels. Smaller scale slab-block towers, SOHO buildings and podiums layer built form and typology to integrate public and private open space. Residential apartment towers, each with varying treatment, have vistas through the site to the hills and mountains beyond. The landscaped arrival plaza to the west is a covered address providing access to the offices, hotel and apartment lifts. An intricate overlay of garden pathways and interconnected ponds and landscape elements, together with existing parks, combine to create a sophisticated network of spaces and a contemporary garden theme. | All buildings are articulated with detailed glass, concrete and metal clad themes, with potential for façade openings, aerial landscape gardens, and screens providing depth, climate protection and garden amenity. Throughout the site, coloured elements are carefully used as a highlight focus and as a way of visually controlling building pattern and scale. | The 260-room, five-star Radisson Blu Hotel, located within the Liuzhou City Plaza development, is articulated as two slender vertical forms. It is detailed with a series of metal perforated screens that de-scale the facade while creating a sense of lightness. Elsewhere glass facades enable clear views to the east and west. The 25,000m^2 hotel includes a 300-seat banquet hall, several restaurants and a recreation centre with pool and fitness centre.

建设地点 中国 柳州
客　　户 柳州阳光壹佰置业有限公司

Location Liuzhou, China
Client Liuzhou Sunshine 100 Real Estate Co. Ltd

雅加达的曼哈顿酒店坐落在雅加达市中心区，是一座125m高、造型优雅的炭灰色塔楼。这座五星级酒店共36层高，有225间标准客房和套房，其中包含一套总统套房、两套豪华套房、102套标准套房和134套高级客房；其它设施包括五间餐厅、一个水疗中心、健身俱乐部、游泳池和一个800座的宴会厅及停车场。| 设计者采用横向的深凹条窗为室内提供遮阳。建筑底部为四面通透的玻璃立方体；位于顶部的立方体则局部突出建筑表面，形成一组富有动态的玻璃盒子。前台设在建筑的顶部的玻璃盒子里。建筑的顶部完全挖空，形成一个巨大的屋顶，其内为一组叠砌的玻璃盒子。

An elegant, 125m charcoal tower located in Kuningan in the heart of Jakarta, the five-star, 36-storey Manhattan Hotel comprises 255 guest rooms and suites – a presidential suite, two deluxe suites, 102 junior suites and 134 superior rooms – and facilities including five restaurants, a spa and fitness club, swimming pool, 800-seat ballroom and car park. | Deep horizontal cuts punctuate the dark, metallic shaft to provide horizontal sun shading. Cubical spaces are fully glazed at the base; at the top they are partially carved out to form a series of dynamic interlocking transparent boxes. The front desk is located in the transparent box on top of the tower. Hotel lobby, meeting rooms and ballroom are stacked together inside a silver box and punched through the glass cube, only to reappear at an angle on the other side.

建设地点　印度尼西亚 雅加达
客　　户　PT Merlynn Park Hotel

Location　Jakarta, Indonesia
Client　　PT Merlynn Park Hotel

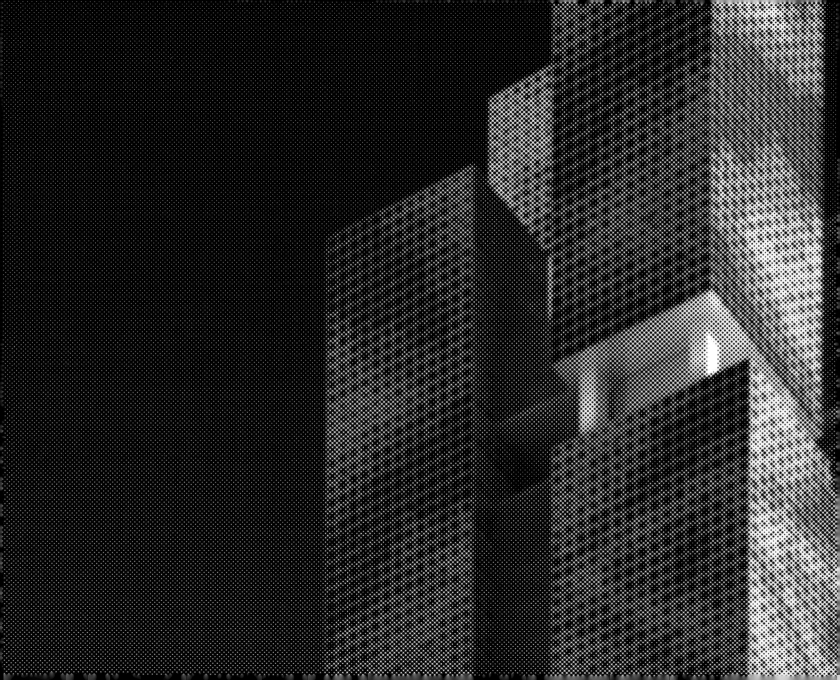

晶簇综合塔
CLUSTER COMPLEX

在国际设计竞赛中，DCM 为迪拜设计了一组用于商业办公和住宅的晶簇综合塔方案，高耸入云的外观是其独创风格的主要特征。非比寻常的是，设计者将塔楼中的各个功能空间在竖向上按序排列——由私人住宅和办公空间至私人社交空间，再到完全向公众开放的空间。| 在景观设计方面，一系列具有说服力的建筑元素不仅与裙楼的设计形成呼应，也适应当地气候的特点。塔楼覆盖了办公区域、酒店和住宅，引人注目的外观与裙楼和景观形成互补。塔楼由四根"立柱"组成，每一根"立柱"是一幢独立的办公楼，这些办公楼沿着高达 255m 的中庭竖向紧密堆砌。壮观的办公塔楼下方是一座座相对较小的酒店和公寓，居住空间距办公和休闲场所很近，出入方便。| 绵延的网状天篷遮蔽裙楼，同时也作为塔楼的绿地。天篷造型多变，创造出类似沙丘或帐篷的具有雕塑感的外观，同时在其下方形成形状各异的空间。裙楼的遮阳板和塔楼的外立面使用金属网板，可以很好地适应当地气候特点，令这个建筑群在迪拜诸多高层建筑项目中显得与众不同。| 公共空间的塑造是设计理念中重要的一部分，体现在景观的统一性、前室的使用以及使用多层门庭结构以增进建筑内部的重要横向联系等方面。这些小空间遍布整个建筑群，提供了沟通和交流的空间——这些社交设施使这个项目几乎成为了一项社区建设。

A striking vertical street characterises the innovative nature of a design competition entry for a precinct of commercial office and residential towers in Dubai. Unusually, activities are vertically layered, progressing in a sequence from private living and working areas, through a privileged interactive zone, to areas where full public access is available. | A series of powerful architectural elements set in a landscape formation form a unique response, not only to the design of a podium, but also to the prevailing climate. The towers, which accommodate the office, hotel and residential users, become striking complementary forms. Comprised of four shafts, each is a series of individual office buildings clustered vertically along a dramatic 255m atrium. Below the imposing office tower sit smaller blocks of hotel and apartment towers. Residents are close to work and recreation, with easy transition from one to the other, and back again. | A continuous mesh canopy envelops the podium, and grounds the towers in a tapestry of green. A sculptural form evoking a sand dune or tent, the canopy's variable profile creates a wide variety of spaces beneath it. The perforated metal, as both podium screen and building façade, performs well in the harsh environment, setting it apart from other high-rise projects in Dubai. | Public space is fundamental to the concept, from the integration of landscape through to the use of atria and multi-level lobbies encouraging vital horizontal connections. These interstitial spaces extend through the entire complex, promoting discourse and interaction – the social vitality to turn it from a project into a community.

建设地点	阿拉伯联合酋长国 迪拜
客　户	Mohammad Al-Mojil 集团
荣获奖项	2011 年 MIPIM 建筑评论未来项目奖
	——高层建筑类最佳建筑奖

Location Dubai, United Arab Emirates
Client Mohammad Al-Mojil Group
Awards MIPIM Architectural Review Future Projects Award
– Winner Tall Buildings Category, 2011

重庆南岸国际新城是一个由28幢塔楼组成的地标式开发项目，坐落于扬子江和嘉陵江的交汇处，遥望重庆渝中半岛，设计任务包括新城总体规划和建筑设计。| 这个新城市中心的设计旨在为日益增长的内城人口提供配套服务设施，并按照合理的分期规划逐步建成。该项目拥有住宅、办公、商业和娱乐空间，总建筑面积达120万 m²，其中公寓数量达到12800套。项目的中心标志是一幢200m高、用于办公和酒店的塔楼。| 在总体规划中，设计者根据场地的坡度，成功地将一个历史保护区整合进规划当中，同时提供了充足的绿地。地块的基本结构由两条互相垂直的轴线——滨江轴线和城市轴线构成，并使建筑群在重庆的城市文脉中脱颖而出。城市轴线指向市中心，沿斜坡向上延伸形成多级广场。两条轴线的交汇处形成一个社区广场。

Masterplan and architectural design of a landmark development of 28 towers that sit where the Yangtze River meets Jialing River, looking across to Yuzhong Peninsula. | The new urban centre is designed with facilities to service the growing inner urban population, delivered according to a rationalised staging plan. The 1,200,000m² living, office, retail and leisure precinct includes 12,800 apartments. A 200m high office and hotel tower is planned as a centrepiece. | Responding to the steep site, the masterplan successfully integrates a historical reservation area along with generous provision for green spaces. Two axes form the basic structure of the site and define it within the broader context of Chongqing. The waterfront axis is perpendicular to the city axis. This land axis runs from the city centre up the escarpment. A terraced plaza becomes the centre point of the development where the axes meet.

建设地点 中国 重庆
客　　户 阳光壹佰房地产开发有限公司

Location Chongqing, China
Client Sunshine 100 Real Estate Co. Ltd

URBAN DESIGN + TR

城市设计 + 城市基础设施

INFRAS-
UCTURE

网桥
WEBB BRIDGE

网桥是墨尔本庞大的步行和自行车交通系统的一个组成部分，也是点缀城市的一颗明珠。这件与艺术家罗伯特·欧文（Robert Owen）合作设计的作品给人以视觉的冲击，为往来的骑车人和行人带来愉悦的体验，也使人们可以沿桥观赏城市的景色。| 网桥由两个独特的部分构成：一个 145m 长的原有铁路桥和一个新建的 80m 长的曲线坡桥。坡桥连接了高度不同的两端，在南岸形成入口。两个部分无缝对接，强调了蜿蜒结构内的体积和容量。| 两个部分形成了统一的雕塑般的外观，新的结构表现出新老之间以及过去和未来之间的新的联系和新的结合。从远处眺望，它是一个功能空间，为通行提供方便。作为一件艺术品，网桥轮廓鲜明，造型优美，轻盈而简洁。空间富有动感和变幻，营造出一种独特的氛围。

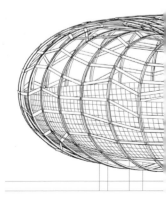

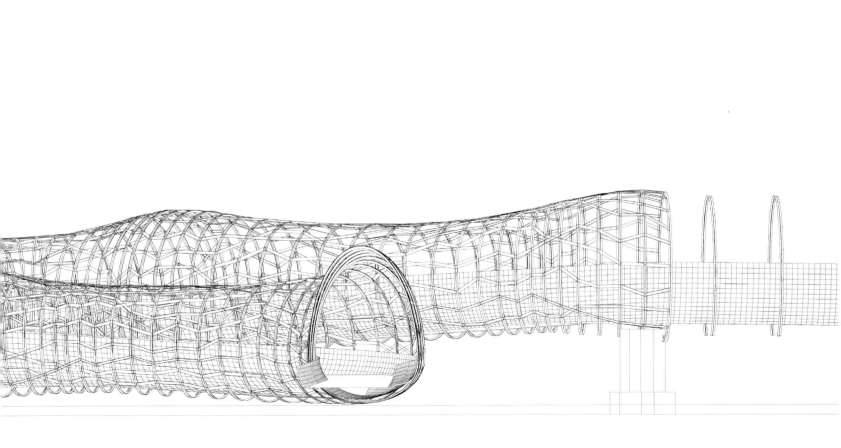

A small jewel-like urban element, the bridge is part of Melbourne's extensive system of bicycle and pedestrian paths. Developed in collaboration with artist Robert Owen, the visually compelling Webb Bridge provides a lively moment for cyclists and pedestrians on the move and allows for leisurely appreciation of the city views experienced along the bridge. | It comprises two distinct sections: the 145m long existing structure and a new, curved, 80m long ramped link. The ramp takes up level changes and creates a point of arrival at the south bank. The two sections are joined seamlessly, with emphasis on volume and containment within the curved and sinuous form. | The two parts become a unified sculptural form. The resulting structure suggests a new connection, or a knot, between the old and new, past and future. From afar, it is perceived as an object that becomes, in turn, a place of action and transition as one uses it. As an object, it appears as a delineated structure, a sensuous volume, light and linear. Space is seen as atmospheric, dynamic and transitional.

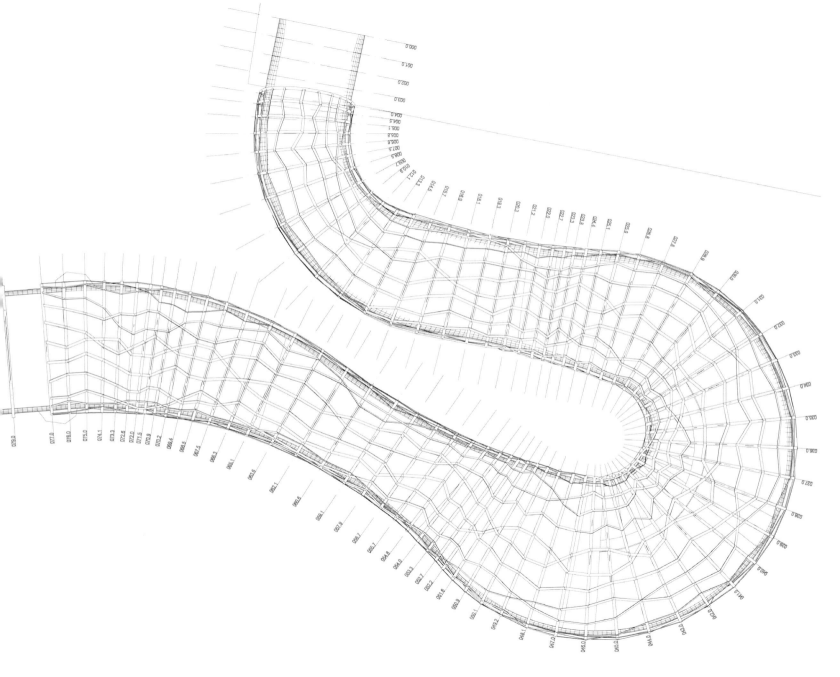

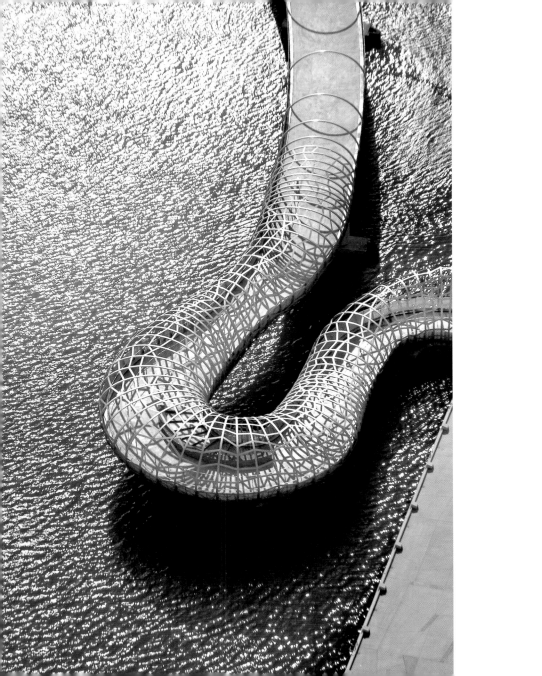

建设地点	澳大利亚 墨尔本
客　户	VicUrban（维多利亚州政府）， Mirvac Development and Docklands Authorities
荣获奖项	－ 2005 年澳大利亚联邦建筑师协会（维多利亚） 　 Joseph Reed 最佳城市设计奖 － 2005 年澳大利亚联邦建筑师协会城市设计奖 － 墨尔本港区公共艺术项目设计邀请赛头奖

Location	Melbourne, Australia
Client	VicUrban (State Government of Victoria) Mirvac Developments and Docklands Authority

Awards	- Joseph Reed Award for Urban Design,
	  Australian Institute of Architects (VIC), 2005
	- National Commendation for Urban Design,
	  Australian Institute of Architects, 2005
	- First prize, limited design competition for a
	  public art project in Melbourne's Docklands

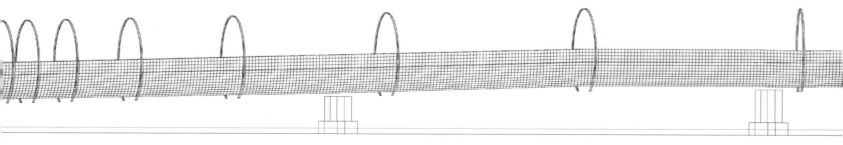

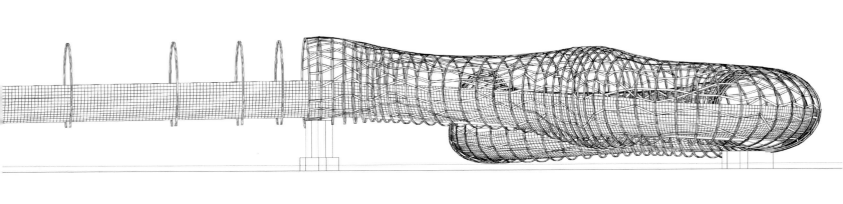

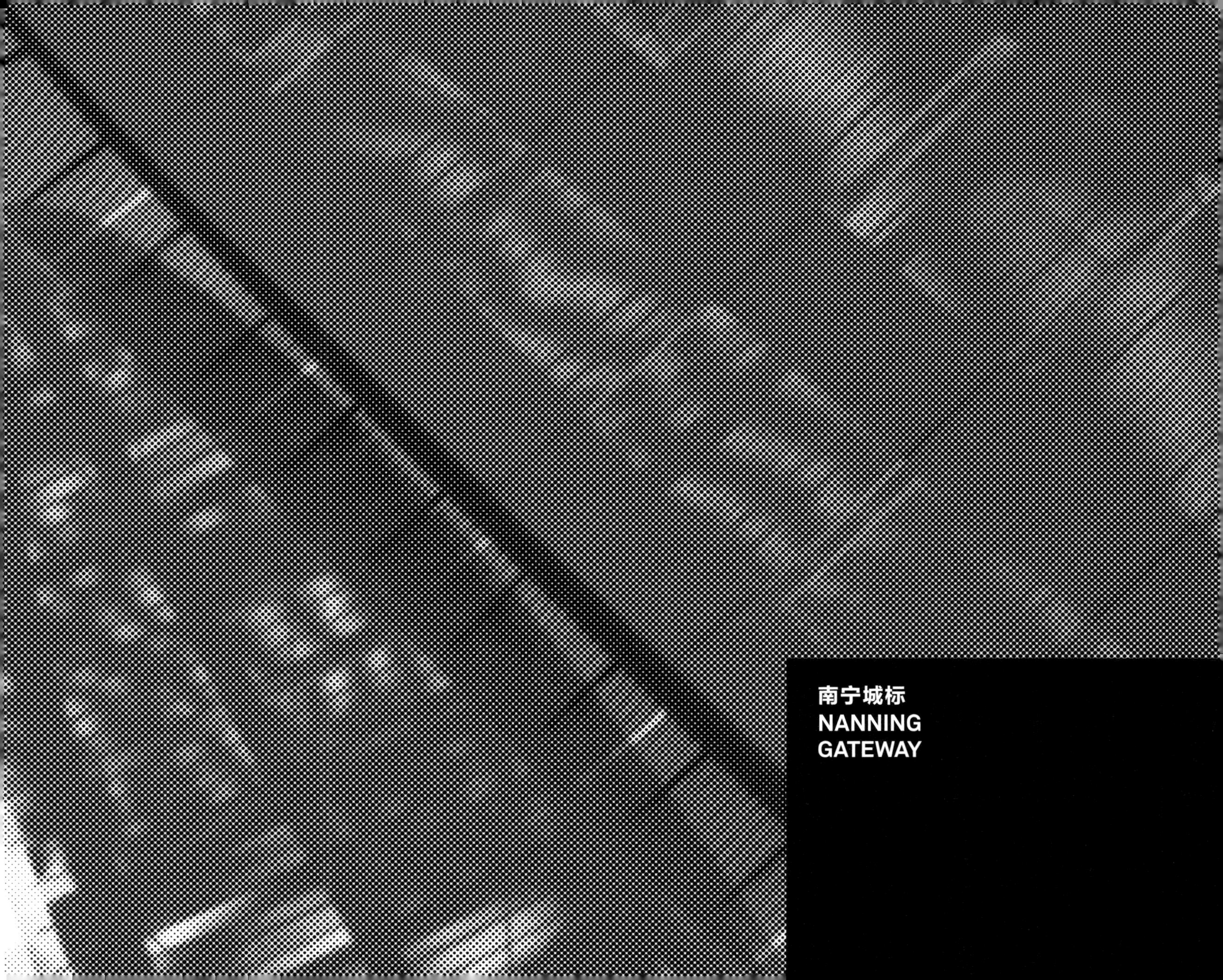

南宁城标
NANNING
GATEWAY

南宁城标的设计理念来源于在大地景观中放置抽象雕塑的概念，它同时还是一件与时间有关的艺术作品。随着车辆的前行，视觉上原本静止的雕塑为经过这里的人们提供动态的时空体验，这一构思发掘出了高速公路上独特的设计契机。| 当人们经过公路收费站出口的时候就可以看到两座巨大的花朵雕塑位于高速公路的侧面。在车辆驶离出口的一瞬间，两朵花中的一朵就开始逐渐绽放；随着车辆的前行，其中一朵渐变成一组花瓣，成为幻象。| 十片花瓣散落在高速公路北面的山坡上，绵延600m，花瓣的高度则根据距离远近的不同，在10m到20m之间变化。

The Nanning Gateway is based on the concept of large abstract sculptural forms in landscape. The physical act of movement transforms an otherwise static sculpture into a dynamic space-time experience, exploiting the unique opportunities available to freeway design. | First viewed while pulling away from the tollgates, the design initially presents as two monumental sculptural flowers flanking the freeway. For one brief moment, they bloom together, but as the traveller proceeds, one of the flowers deconstructs into a series of singular petal elements, revealing itself as an optical illusion. | A total of 10 petals are dispersed across the sloping banked north side of the road, over a distance of approximately 600m. Owing to the distance between the nearest and farthest petals, their sizes vary significantly from 10m to 20m tall.

建设地点 中国 南宁
客　　户 南宁规划管理局

Location Nanning, China
Client Nanning Administration Bureau of Planning

十六铺南滨
总体规划研究
SHANGHAI SHI LIU PU
SOUTH BUND
MASTERPLAN

该项目受上海土地部门的委托,以国际设计邀请赛的方式,开展对上海市区内最具价值同时也是最敏感的区域之一的研究,为地块未来开发提供纲领性的发展规划。| 总体规划将景观、行人活动、机动车出入口以及轮渡、未来地铁和公车线路全部纳入考虑。这项规划的战略性目的包括优化和强化地块与水滨和公园公共空间之间的联系,改善黄浦江和内陆之间视觉和空间连接,统一协调滨水空间,制定合理的开发强度以及创建一个地标式开发区域。| 研究工作的首要目标是为该地块确定一个合适的容积率——使之既符合环保要求和合理开发强度,又能够很好地和城市大环境融为一体。然而,最大的挑战还是来自于地块本身所具有的重要文化价值。成果需要将"老上海"(中国传统老城区和 19 世纪"现代"上海)和"新上海"的城市特色有机地结合起来,从而有效处理黄浦江与地块间的滨江大道,并修建完善的交通系统使人们可方便前往地块内的高端办公楼、酒店和精品商业区。研究同时也设计了便利的交通来连接老城区和新滨江海洋公园以及游轮渡口。地块北端将设立一个地下公交车枢纽站,上面是城市开放空间和公园。| 研究工作分两期完成。第一阶段是调查地块及周围环境,以形成基本的策略和概念。设计概念提交给专家委员会进行详细地讨论,以便确定第二阶段的工作。二期研究工作的内容主要是按照专家评审意见和建议深化概念设计。二期研究结果显示,合理的开发建筑面积应控制在 29 万 m^2。

A study of one of the most sensitive areas in Shanghai to provide guidelines for future development. This project was commissioned by Shanghai Land and required a study of one of the most valuable and sensitive areas in Shanghai with the objective of providing guidelines for future development. The study was undertaken as a limited international design competition. | The masterplan considered view, pedestrian movement, vehicular access and transportation including ferry, future metro and bus routes. Its strategic objectives were to optimise key connection to the waterfront and public park open space systems, manage and improve physical and visual connections between the river and inland zones, enhance continuity of the waterfront, both visually and physically and create a landmark development within the city. The study's first aim was to establish a plot ratio appropriate to the site – sustainable and well integrated into the broader context of the city; however the greatest challenge came from the site's cultural significance. The outcome integrated the old (both Chinese traditional old downtown area and 'modern' 19th century Shanghai) and new in built form, effectively managing the thoroughfare between the site and the river and providing seamless access to the site which contains high-end offices, hotels and boutique retail. It provides links between the old downtown area to the new waterfront marina recreation park and cruise docks. An underground bus depot/hub was also located at the north end of the site with open urban space/parkland above. | The study was undertaken in two stages. The first was to investigate the site and its context, developing the general strategy and concept. The concept was presented to the expert panels and discussed in detail to set up the second stage of work, which was to develop the concept generated from the first stage incorporating expert comments and suggestions. This resulted in a proposed floor area above ground of 290,000m^2.

建设地点 中国 上海
客　　户 上海地产（集团）有限公司（上海市土地储备中心）

Location Shanghai, China
Client Shanghai Land (Group) Co. Ltd

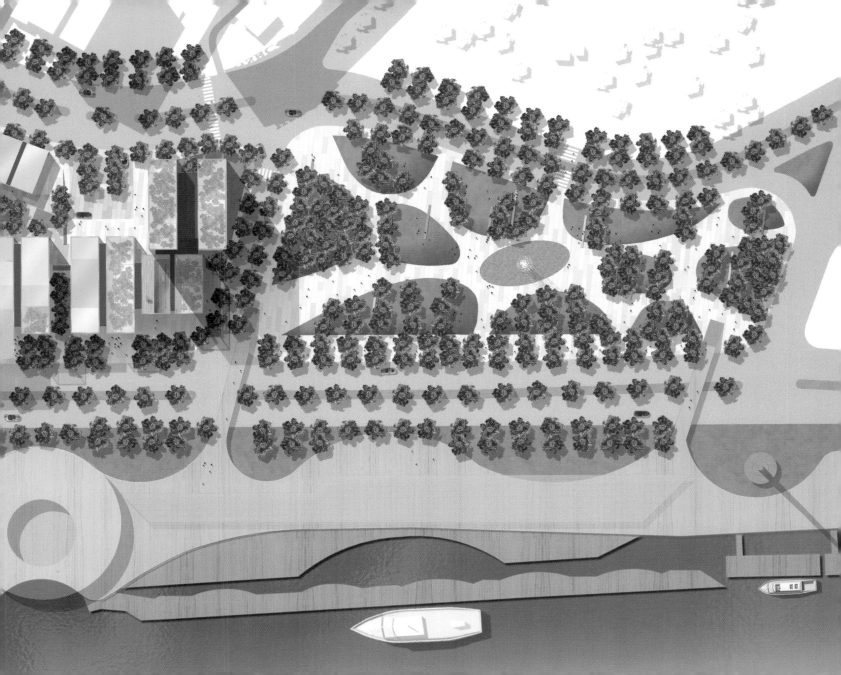

金沙湖
步行景观桥
国际竞赛
JINSHA BRIDGE
COMPETITION

金沙湖步行景观桥是 DCM 在国际竞赛中的参赛作品,它坐落在杭州新城的金沙湖上。| 步行景观桥鲜明、优雅的外观生动地诠释了地方传统文化内容。人行桥由三根彩带——低桥(通行桥)、高桥(休闲观景桥)和拱形支撑结构有机地组成一体,在结构和视觉效果上都和谐共存。| 通过对不同路径的设计,桥梁能为不同目的的行人带来不同的精彩体验。通行桥是到达湖对岸的最直接的通道,从外观上看它支撑着整个桥体,由岸边缓缓升起,在航道上方形成一个宽平的拱形。高桥长 566m,在湖岸处渐渐升起,与低桥呼应,但很快就展示出自己的个性。它一路蜿蜒,起伏的观景桥使游人可以在穿行湖面的过程中眺望到金沙湖新区的全景。拱形支撑结构如正弦曲线在湖面波动,曲线的振幅渐增,在湖心位置达到最大,并与低桥和高桥在空中交错有致。拱桥在湖的上空形成一条优美的抛物线,曲线最高点距离湖面 51m,体现了其固有的结构完整性。| 在蓝天的映衬下,金沙桥在四周建筑中尤为夺目。它修长的结构和开放的空间便于游人观景,阳光可以在桥体和湖面上形成变幻的光影,动感的外观带给人视觉的盛宴,像一场流畅而优美的舞蹈演出。

Denton Corker Marshall's entry in a limited international design competition, since abandoned, for a landmark pedestrian bridge in the expansion zone of the historic city of Hangzhou – Jinsha New Town. | The distinctive, elegant form energetically interprets traditional references. Three ribbon elements – a lower deck, upper deck and arched support structure – combine into an integrated object, in harmony both structurally and visually. | The design delivers surprise and delight for pedestrians and cyclists on two dedicated routes. The most direct path across the lake, the lower deck visually anchors the composition, growing gently from shoreline to rise above the navigation channel in a broad, shallow arch. The 566m upper deck echoes the lower deck as it sets out from the shore, but quickly displays its own character. Twisting from one side to the other, the elevated leisurely route affords panoramic views across the lake to New Hangzhou. | The Arched Support Structure flows in a sine-wave of increasing intensity, engaging with the lower and upper decks at regular intervals. A sequence of curves peaks 51m above the lake in a graceful parabolic arch, delivering inherent structural integrity. | Rising into the sky, the bridge attains a presence amid the scale of surrounding buildings, and its slender structure and open spaces allow views and light to penetrate. The dynamic form is visually complex and stimulating; a flowing and elegant choreography.

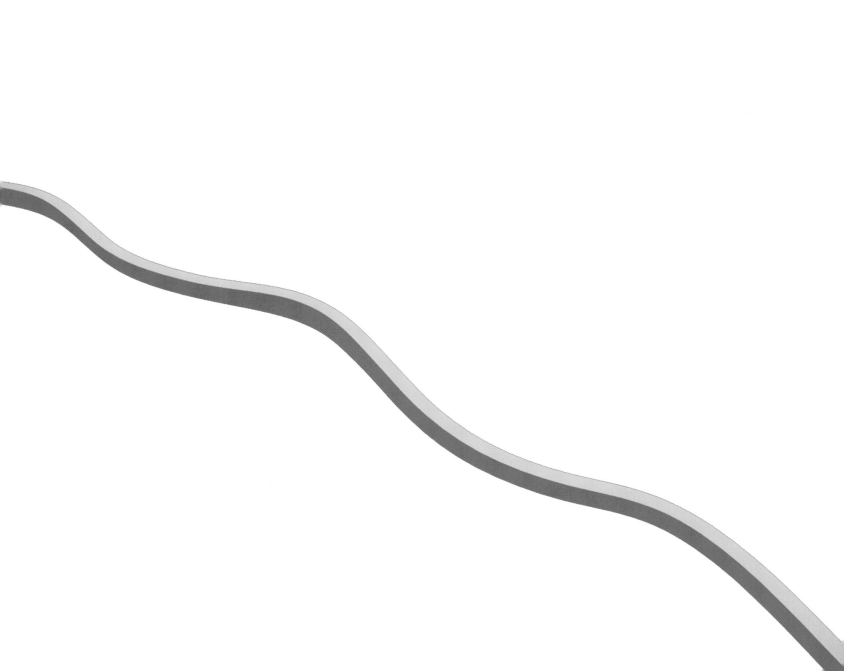

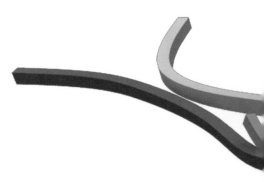

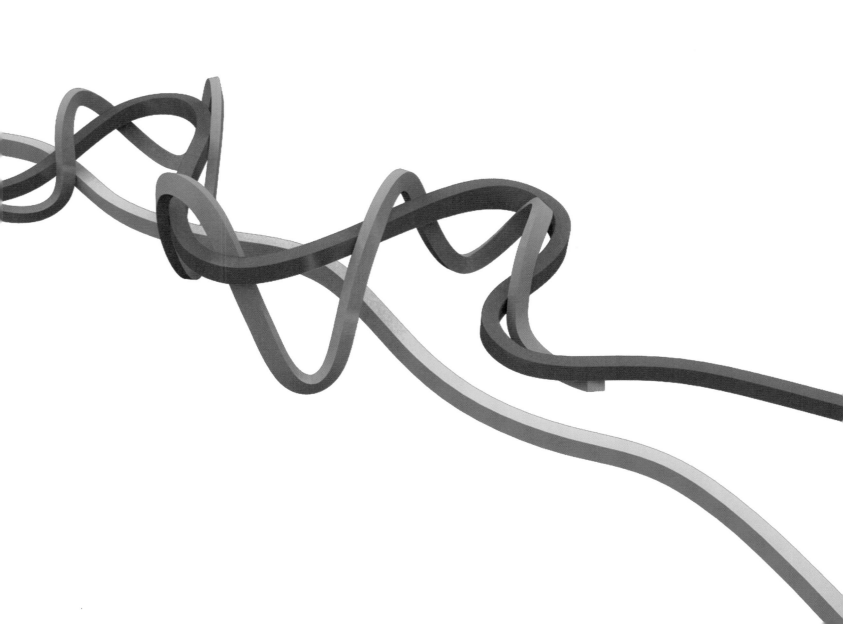

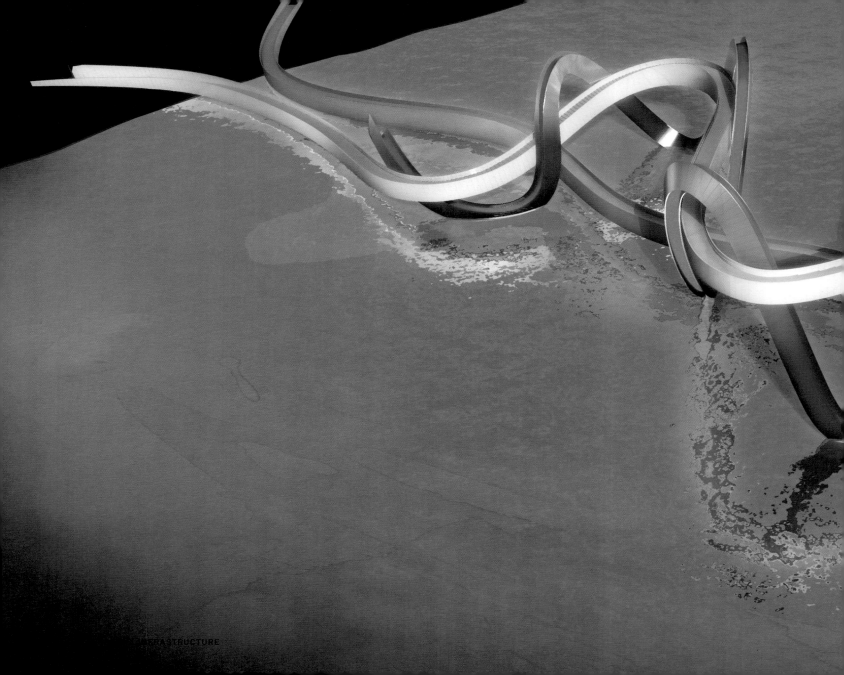

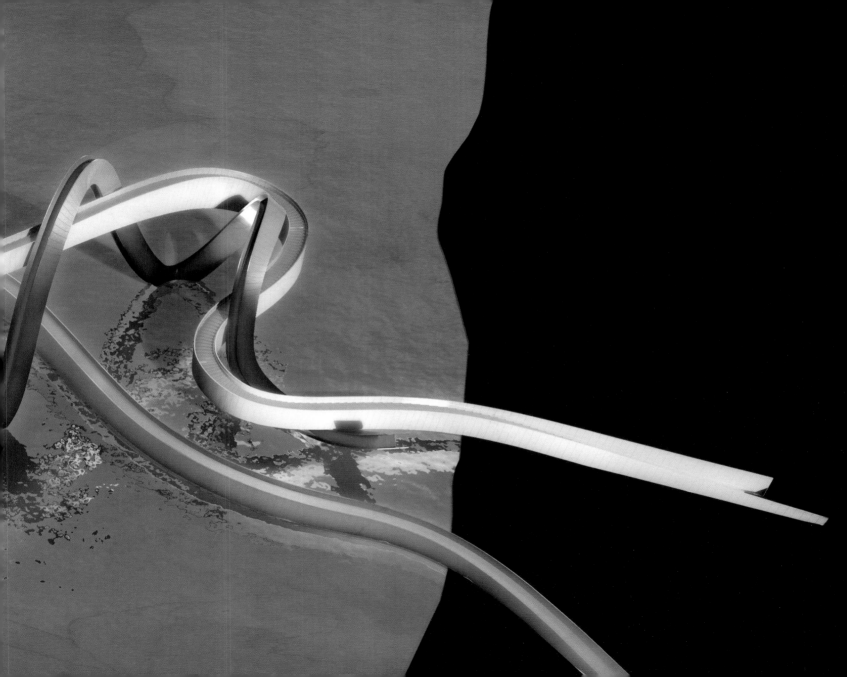

建设地点 中国 杭州
客　　户 金沙湖投资发展有限公司

Location Hangzhou, China
Client   Jinsha Lake Investment & Development Co. Ltd

作者
THE AUTHORS

朱剑飞 副教授　朱剑飞是墨尔本大学建筑建设规划学院副教授。主要著作有 Chinese Spatial Strategies（Routledge，2004 年）和 Architecture of Modern China（Routledge，2009 年），并主编了《中国建筑 60 年（1949—2009）：历史理论研究》（中国建筑工业出版社，2009 年）。

聂建鑫 先生　聂建鑫曾于 2002 年撰写了《澳大利亚 DCM 作品实录》一书，由中国建筑工业出版社出版。聂建鑫先生目前在北京工作，他是澳大利亚注册建筑师，澳大利亚联邦建筑师协会会员以及澳大利亚 SDG 设计集团合伙人。

Associate Professor **Jianfei Zhu**　Jianfei Zhu is Associate Professor of the Faculty of Architecture, Building and Planning, at The University of Melbourne. He is the author of *Chinese Spatial Strategies* (Routledge 2004) and *Architecture of Modern China* (Routledge 2009), and the editor of *Sixty Years of Chinese Architecture: History Theory and Criticism* (CABP 2009).

Mr **Jian-Xin Nie**　Jian-Xin Nie was the author of *Denton Corker Marshall* published in 2002. He is based in Beijing, China and is also an Australian registered architect, a member of The Australian Institute of Architects and a Partner of SDG Shine Design Group, Australia.

丹顿·廓克·马修建筑设计事务所向所有员工和参加书中项目设计工作的建筑师致以诚挚的谢意。尤其要向公司的高级副总经理和副总经理致以谢意，感谢他们多年来始终如一为丹顿·廓克·马修所作的贡献。

墨尔本 | **董事** | John Denton, Barrie Marshall, Adrian Fitzgerald, Ian White, Neil Bourne, 龚耕 , Peter Williams, Wojciech Pluta, Anna Piatkowska | **顾问** | Bill Corker | **高级副总经理** | Nicola Smith, Noel Tighe | **副总经理** | Anthony Blazquez, Mark Bol, Anne Clisby, Eric Schatz | **伦敦** | **董事** | Stephen Quinlan, John Rintoul | **副总经理** | Angela Dapper, Phillip Millership, Robert Holford | **曼彻斯特** | **高级副总经理** | Irwin Lopez | **雅加达** | **董事** | Budiman Hendropurnomo | **副董事** | Dicky Hendrasto | **副总经理** | Farida Utari, Albertus Sutianto, Fajar Setiawan | **园林部经理** | Fauzi Wahyudin

摄影

Tim Griffith, John Gollings, Shannon McGrath, Nicholas Lee
Katarina Stube, Blain Crellin, Eric Sierens, William Furness, Yori Antar

合作公司 / 事务所

威尼斯双年展澳大利亚馆：	与 Fare Architects 合作设计
怀特艾利沙医学研究院：	与 S2F 合作设计
新加坡亚洲广场：	与新加坡 61 建筑设计有限公司合作设计
北京乐天银泰百货：	与北京建筑设计研究院合作设计
北京中坤大厦：	与东方筑中建筑设计事务所合作设计.
中国北京延庆县落樱酒庄：	与中科院建筑设计研究院（北京）合作设计
南宁欧景城市广场：	与广西华蓝设计集团有限公司合作设计
柳州城市广场和丽笙酒店：	与广西华蓝设计集团有限公司合作设计
重庆南岸国际新城：	与中国建筑科学研究院中国建筑技术集团有限公司重庆分公司建筑设计院合作设计
墨尔本网桥：	与艺术家罗伯特·欧文合作设计

鸣谢
ACKNOWLEDGEMENTS

Denton Corker Marshall would like to acknowledge and thank the many people who have worked for the practice and those who participated in the projects illustrated in this book. In particular, the Associates and Senior Associates who have made a long-standing, consistent contribution; Denton Corker Marshall would like to thank them for their dedication.

Melbourne | *Directors* | John Denton, Barrie Marshall, Adrian FitzGerald, Ian White, Neil Bourne, Greg Gong, Peter Williams, Wojciech Pluta, Anna Piatkowska | *Consultant* | Bill Corker | *Senior Associates* | Nicola Smith, Noel Tighe | *Associates* | Anthony Blazquez, Mark Bol, Anne Clisby, Eric Schatz | **London** | *Directors* | Stephen Quinlan, John Rintoul | *Associates* | Angela Dapper, Phillip Millership, Robert Holford | **Manchester** | *Senior Associate* | Irwin Lopez | **Jakarta** | *Director* | Budiman Hendropurnomo | *Associate Director* | Dicky Hendrasto | *Associates* | Farida Utari, Albertus Sutianto, Fajar Setiawan | *Landscape Associate* | Fauzi Wahyudin

Photography
Tim Griffith, John Gollings, Shannon McGrath, Nicholas Lee
Katarina Stube, Blain Crellin, Eric Sierens, William Furness, Yori Antar

Architects in Association
- The Australian Pavilion, Venice in association with Fare Architects in Rome, Italy.
- The Walter and Eliza Hall Institute of Medical Research, Melbourne in association with S2F.
- Asia Square, Singapore in association with Architects 61 as Architect of Record.
- InTime Lotte Shopping Centre, Beijing in association with the Beijing Institute of Architecture Design.
- Zhongkun Tower, Beijing in association with Oriental Elite Design Institute.
- Zhongkun Vineyard, Yanqing County in association with the Institute of Architecture Design and Research, Chinese Academy of Sciences.
- Euro-City Plaza, Nanning in association with Guangxi Hualan Design Group Co. Ltd.
- Liuzhou City Plaza + Radisson Hotel, Liuzhou in association with Guangxi Hualan Design Group Co. Ltd.
- Southbank International New Town, Chongqing, in association with China Building Technology Group Co. Ltd (Chongqing), Architectural Design Office, China Academy of Building Research (CABR).
- Webb Bridge, Melbourne in association with artist, Robert Owen.

图书在版编目（CIP）数据

筑作 WORKS：丹顿·廓克·马修建筑设计事务所：汉英对照 / 朱剑飞，聂建鑫主编．
——上海：同济大学出版社，2013.10

ISBN 978-7-5608-5295-9

Ⅰ．①筑⋯
Ⅱ．①朱⋯ ②聂⋯
Ⅲ．①建筑设计－作品集－世界－现代
Ⅳ．① TU206

中国版本图书馆 CIP 数据核字 (2013) 第 222178 号

筑作 WORKS：丹顿·廓克·马修建筑设计事务所	Denton Corker Marshall	朱剑飞 聂建鑫 主编

责任编辑	江 岱	经 销	全国各地新华书店
助理编辑	陈 淳	印 刷	精一印刷（深圳）有限公司
责任校对	徐春莲	开 本	787mm×1092mm 1/16
装帧设计	唐天辰	印 张	31.75
出版发行	同济大学出版社	印 数	1—2 500
	www.tongjipress.com.cn	字 数	792 000
	地址：上海市四平路 1239 号	版 次	2013 年 10 月第 1 版　2013 年 10 月第 1 次印刷
	邮编 200092	书 号	ISBN 978-7-5608-5295-9
	电话 021-65985622	定 价	320.00 元